How to Make the Most of Your Mind

Using Your Brain, Body and Environment to Improve the Quality of Your Life

Large Print

William J. Winslade

and Heber Taylor

Thompson and Treanor, Publishers

T&T

Galveston, Texas

T&T

ISBN: 978-1-7339050-3-9

Dedication

To Melvyn Schreiber, M.D.

A physician and professor of
medicine, a perennial student,
lifelong learner and tireless reader.
Both authors have benefitted
greatly from his wisdom and
friendship.

CONTENTS

Part II: **INTERDISCLIPLARY LEARNING**

Part III: IMPROVING EDUCATION

INTRODUCTION

About this book

If you are reading this sentence, you are conscious. We humans are conscious unless we are in a deep sleep or coma or are anesthetized or suffering from a medical condition, such as an injury that disrupts brain function. When we are conscious, we can think, talk, make plans and engage in all kinds of activities — the very things that make us human.

But what is consciousness? We know that our brains are essential to our conscious life, that our bodies and the environment influence and impact our consciousness.

Scientists and researchers have studied brains — living and dead — to try to understand how they work. But a full and deep understanding of consciousness has proven to be challenging and elusive.

The human brain is perhaps the most complicated thing in the universe. Don't we all wonder how the brain works, and wouldn't we like to know more?

This book began when William J. Winslade, a professor with The University of Texas Medical Branch's Institute for the Medical Humanities, decided he did, in fact, want to know more. Winslade, who had written a book called *Confronting Traumatic Brain Injury*, published by Yale University Press, decided that the public should have a much broader understanding of advances in the neurosciences. Thus began the Robert and Russell Moody Lecture Series, "Consciousness and the Birth, Life and Death of the Brain."

Each semester for three and a-half years, a couple of speakers — scientists, physicians, philosophers, psychologists and other thinkers — came to Galveston to explain new developments in neuroscience in language that all of us could understand.

The first lecture, in October 2016, featured Christoff Koch, president and chief scientific officer at the Allen Institute for Brain Science in Seattle, perhaps the most famous brain scientist today. The last, in March 2019, featured Norman Doidge, a physician and author of the best-selling books *The Brain that Changes Itself* and *The Brain's Way of Healing.*

From the beginning, the whole idea was for the lectures to be public. Some of the speakers also spoke to faculty and students at the University of Texas Medical Branch, the state's oldest medical school. But the public lectures were designed to reach a broad audience, and they did.

Each of us has a brain, and the experts who spoke in the lecture series gave the public a broad view of the research in simple English. This book seeks to pass that information on to a larger audience.

The book is in three parts:

Part I summarizes what the lecturers said about the lifecycle of the brain. You'll learn something about the way the brain works, how to take care of it, and how to get the most out of it.

Part II is about learning and about what the new science of the brain suggests we can do to make the most of minds. It makes a case for an approach to learning that cuts across the boundaries of the usual academic disciplines.

Part III is a series of contributed chapters from people in education who provide a more detailed look at how interdisciplinary learning could change our schools.

The appendices are longer than usual and offer another way to understand the work of the experts who talked about their research. Before every lecture, *The*

Galveston County Daily News published articles, including Q&As with the speakers and commentary by Professor Winslade. Those pieces, slightly edited, are published here with permission of *The Galveston County Daily News.* A reader who has difficulty in understanding a point about the brain might get additional help from those articles.

All of us have a brain. This book is an opportunity to learn more about how yours works and how you can make the most of it.

PART I

THE BIRTH, LIFE AND DEATH OF THE BRAIN

Chapter 1

The development of the brain

It would be a good thing, perhaps a wonderful thing, if the brain came with an owner's manual. It does not, of course, but advances across many fields — including neuroscience, philosophy, imaging technology and psychology — are giving us a better picture of how our brains develop and work. It's also becoming clear that our minds emerge from the brain, the body and the environment.

It's obvious that if the brain dies, the mind dies with it. But it's also obvious that when we are talking about our minds, we are talking about our bodies' capacity to provide sensations to our minds. We learn by experiencing the world, and those experiences include emotions and thoughts, as well as sensations. Our minds also are influenced by environment. This book is part of your environment. In reading it, you are taking in influences from your environment. You are getting information on the research and thinking of some leading authorities in the field. The countless conversations and experiences with parents, teachers and friends also shape your mind.

But it all starts with the brain. Everyone who has a brain should know a little bit about how it works. Here's an overview of how a brain develops, from birth to maturity.

Before birth

When does "the magical journey of consciousness begin?"

Christof Koch posed that question in an article in *Scientific American*. In forming an answer, Koch described the beginning as a "road to awareness," rather than one specific event at a singular moment in time.

The sensory systems — including those for sight and hearing — develop at different rates. Ears develop around the 20th week, and the fetus is able to process external sounds at around the 25th week. By contrast, while visual development begins around the 20th week, a child's visual system is not fully developed at birth. Infants see blobs, rather than sharp images.

The physical structures that support consciousness, the thalamocortical complex, develop between the 24th and 28th weeks. Around this time, the first signs of large-scale electrical activity across the brain can be recorded and intermittent episodes of electroencephalographic rhythms can be detected. "Thus, many of the circuit elements necessary for consciousness are in place by the third trimester," Koch says.

At this stage, most premature babies do, with proper care, survive. They are much easier to study for interested doctors than a fetus at the same stage of development.

Koch suggests that it's a mistake to equate the fetus with a preterm baby of the same gestational age, to consider it "an unborn newborn." That view ignores the uterine environment. As Koch points out, the fetus is suspended in a warm, dark cave, connected by the umbilical cord to the mother's placenta, which provides it with everything it needs to survive and grow.

The fetus, in the third trimester, is almost always asleep.

Using electroencephalography, we can observe two distinct states. Active sleep is distinguished by breathing, swallowing, licking and moving the eyes. Quiet sleep features no breathing or eye movements. The two stages of sleep likely correspond to the two types of sleep — rapid eye movement and slow wave or deep sleep — common to all mammals after birth.

"What is fascinating is the discovery that the fetus is actively sedated by the low oxygen pressure (equivalent to that at the top of Mount Everest), the warm and cushioned uterine environment and a range of neuroinhibitory and sleep-inducing substances produced by the placenta and the fetus itself. …" Koch says. "From all this evidence, neonatologists conclude that the fetus is asleep while its brain matures."

When teenagers and adults go to sleep, they frequently report dreams, a form of consciousness in which we encounter friends and loved ones, see and hear, and so on. Dreams are as real as life itself.

"What would dreaming be like for an organism that spends its time suspended in a sort of isolation tank, with no memories, and no way to imagine anything at all?" Koch asked. "I wager that the fetus experiences nothing in utero; that it feels the way we do when we are in a deep, dreamless sleep."

One feature of consciousness that emerges long after birth is self-consciousness. Babies simply are not self-aware. A baby might, in kicking the crib, become aware of pain in her foot. But she would not be experiencing angst at the long-term psychological effects of having suffered pain as an infant.

In summary, if there is consciousness before birth, it most likely does not amount to much. By contrast, natural childbirth is a dramatic awakening.

The brain at birth

Koch says that Nature has

devised the ultimate wake-up call.

"The fetus is forced from its paradisiac existence in the protected, aqueous and warm womb into a hostile, aerial and cold world that assaults its senses with utterly foreign sounds, smells and sights, a highly stressful event," Koch says, noting that, at the same time, the newborn gets a monumental surge of norepinephrine.

As Koch puts it, the infant, "draws its first breath, wakes up and begins to experience life." (Christof Koch, "When Does Consciousness Arise in Human Babies?" *Scientific American*, Sept. 1, 2009)

Infants and toddlers

The experiences children have in the first 18 months of life are vital. They affect not only their development in terms of learning — they affect the development of the brain itself.

Susan Landry, director and founder of the Children's Learning Institute at the University of Texas Health Science Center at Houston, is making an effort to convince parents, child care operators and elected officials of that fact. She contends that far too many children are not reaching their potential because they are not getting adequate stimulation and support in the first three years of life. And, within those first three years, people often overlook the importance of the first 18 months.

"If we wait until pre-K to address developmental problems that are the result of low levels of responsive parenting, including poor cognitive stimulation, then it is often impossible to totally ameliorate learning delays and social and emotional deficits that can have devastating effects for children's outcomes," she said.

"There are solutions, but they require funding, persistence and a commitment from many."

Most children in the United States are not ready to begin kindergarten. They lack the foundational skills, such as being able to count and identify letters, as well as regulate their behavior.

In trying to change this picture, Landry said it's vital to confront this fact: Early experiences are vital in light of the dramatic changes in the brain in the first three years of life.

During that critical period, synaptic connections become increasingly complex. At the same time, neural circuits are developed and become stronger — depending on whether they are used. Stimulation by others, especially by parents or caregivers, is vital. Early experiences determine whether particular connections are dropped because they have not been stimulated. As Landry puts it, experiences with caregivers determine whether the architecture of a child's brain is sturdy or fragile. If the child is not stimulated and supported and fails to develop properly, that deficit becomes increasingly difficult to correct.

Socioeconomic status

A lot of research on development in early childhood has focused on socioeconomic status. The evidence is overwhelming that the effects of poverty are devastating.

Many of these studies have focused on what is missing from the caregiving environment. In general, poor children get less across the board: less responsiveness from adults, less cognitive stimulation through language, less in terms of available materials, such as books, vacations and trips to the museum. That's not true for every child. But, on average, the child born into a low-socioeconomic-status household starts with significant disadvantages.

Studies using electroencephalogram technology show those disadvantages affect the development of the brain. At six months, an age at which we don't have educational assessments, socioeconomic status is strongly associated with higher-power brain waves in regions associated with language development at later ages. Even before language is acquired, the neural networks that allow us to acquire language are slow to develop in some children. The evidence suggests it's simply a lack of stimulation.

Also, a series of studies has established that "hostile" parenting — some parents harshly criticize or strike their children — and family stress negatively affect the child's hippocampus, which is important to the child's memory and his or her ability to manage stress. It's a downward spiral. The more a child is exposed to stress, the weaker the child's ability to manage stress.

The research is clear on what can go wrong in the development of a young child. Researchers have found that the prolonged absence of contact between a child and his or her caretaker may be more damaging to brain development than trauma from physical abuse. Far more children suffer from lack of stimulation than from abuse.

Responsiveness — the ability of an adult to notice and correctly interpret a child's signals and then sensitively engage the child — is the single big predictor in what will happen to the child. Landry says that it is impossible to overstate the importance of responsive parents, caregivers and teachers.

A child's exposure to this type of interaction with a responsive adult predicts all aspects of his or her development and predicts it long-term. Children who get responsive support do better with language, develop emotional and social skills and learn to control themselves.

Language development

The lack of a responsive environment has tragic consequences. One of the greatest problems is the effect on language skills.

It's common for children from low-socioeconomic-status homes who have not had responsive caretakers to enter Pre-Kindergarten programs with developmental delays of 8 to 12 months. The basic problem is that the caregiver's *quality* and *quantity* of language tend to vary with socioeconomic status. From the beginning, there is simply less speech, that is, less discussion in lower-income homes. The quality of speech is also lower. Educated adults tend to be more expressive in the range of their vocabulary. By contrast, adults with less education tend to just ask, "What's this?" or "What's that?" rather than naming things. The child tends to learn "this" and "that," rather than a full range of vocabulary, a phenomenon that researchers call "empty language."

Beginning in infancy, a child from a poorer home often hears less language and poorer quality language. The poor stimulation undermines the child's ability to process speech sounds.

Problems with processing sounds at an early age have a cascading effect on language skills later on. In 1995, Betty Hart and Todd R. Risley published the study that put the phrase "30-million-word gap" into currency (Hart, B., & Risley, T. R., *Meaningful Differences in the Everyday Experience of Young American Children*, Baltimore, MD: Paul H. Brookes Publishing Company, 1995). The study found that, by age 4, low-income children have heard, on average, 30 million fewer words than some of their peers. All over the country, initiatives were launched to close the 30-million-word gap.

What adults can do

One obvious solution is to read to children. When children hear stories, brain areas supporting mental imagery and narrative comprehension are activated. If reading becomes a habit, those parts of the brain function more efficiently.

But further research suggests that reading is only part of the answer. Reading to a child is something done *to* the child, and a more nuanced view of learning includes some give and take.

Researchers use an analogy from tennis: serve and return.

A baby "serves" by communicating interest — pointing or cooing.

The parent or caregiver "returns" the serve by quickly responding, showing interest, using words, gestures and touch to communicate with and encourage the child.

Where that kind of responsiveness happens, you can predict reading success, Landry says. Conversely, unresponsiveness — indifference or neglect — can be devastating. As Landry observed, if you wait for pre-Kindergarten or Kindergarten to intervene, it can be too late. By age 4, it's very difficult to catch up.

Interventions and policy

Some parents and caregivers are models of responsiveness. Their own parents and caregivers were responsive. But that's not universally the case, and Landry has been developing programs to prevent the damage so often done by poverty.

One promising program, Peer Assisted Learning Strategies, used home visits by coaches who taught responsiveness skills to parents. Such programs are expensive, however, and a full-blown national effort just hasn't gotten off the ground. Researchers have focused on a hybrid, in which parents are given an iPad with the training sessions on it. Typically, parents use video technology to record a session of themselves trying to do something with a child. They download the video and contact the coach through Skype for a critique.

The home is not the only

influence on young children who spend time in child care. The quality of that care varies dramatically in terms of responsiveness to the child. Whether in the home or in a daycare center, the keys are the same: development starts with interaction with a responsive adult and exposure to language.

Moral sense

As their brains develop, young children develop capacities of the mind. They develop not only their intellectual capacity, but also a moral sense. When, exactly, does that occur?

Paul Bloom, a Yale psychologist, talks about studies conducted at the Cognition Center at Yale University, led by Karen Wynn, who runs the lab, and Kiley Hamlin. The studies involved toddlers that were too young to talk but who were old enough to watch a puppet show. The puppets were playing with a ball, and most of the puppets played fairly, passing the ball back and forth — sharing it. But one did not. The toddlers not only withheld treats from the "bad" puppet, but one of them gave him a whack on the head.

Videos of the study should give pause to those who have accepted the view that morality depends on the ability to give reasons for what we do. The children who were part of the study were too young to talk, much less articulate reasons. But Bloom argues that they also clearly have something that most of us would call a moral sense.

"These studies suggest that even babies can tell the difference between good acts and bad acts and respond accordingly," Bloom said. "More generally, the results from our labs and others suggest that a lot of morality is bred in the bone — we are born with concern and compassion, the capacity to judge the actions of others, and even a rudimentary understanding of justice and fairness.

"I should add, though, that this isn't the whole story — there is a nastier side to our natures as well. There's a lot of evidence that even the youngest babies carve the world into Us versus Them — and they are strongly biased to favor the Us. We are very tribal beings. Our natures are not just kind; they are also cruel and selfish. We favor those who look like us and are naturally cold-blooded towards strangers. And so learning and socialization are critically important if we want to transform babies into fully moral beings"("Paul Bloom set to Speak Nov. 2," *The Galveston County Daily News*, Oct. 25, 2016).

Question of value

To get a fuller understanding of Bloom's views, it's helpful to follow his interest in Edward L. Thorndike, a psychologist who, during the Great Depression, conducted research on values. Thorndike asked people: How much would I have to pay you to ... ? For example, he asked, How much would I have to pay you to remove one of your front teeth? The average answer, adjusted to today's dollars, was $74,000. When Thorndike asked how much he'd have to pay to get a person to kill a cat, the average answer was $164,000.

Bloom argues that the study suggests that humans are generally more reluctant to do something morally wrong than to do something that disfigures them.

Why is that so?

Bloom says that Adam Smith, one of the leaders of the Scottish Enlightenment, had his finger on the answer. Smith made an extraordinary claim about human nature in his book *The Theory of Moral Sentiments*.

"How selfish soever man may be supposed, there are evidently some principles in his nature, which interest him in the fortune of others, and render their happiness necessary to him, though he derives nothing from it except the pleasure of seeing it. … That we often derive sorrow from the sorrow of others is a matter of fact too obvious to require any instances to prove it."

Smith, who is better known as the author of *The Wealth of Nations*, is often seen as arguing that selfishness — at least in acting through self-interest in a society with free markets — can produce good. In fact, he argued that it's a part of human nature to act in the interests of others.

Bloom says that claim raises

developmental questions: Where does that tendency toward altruism come from? Does it come from our nature or from our nurture? Are we savages or something else?

Many Enlightenment thinkers agreed with Smith that the moral sense was an innate, defining trait of humans. It's an interesting claim to investigate in light of evolutionary biology and developmental psychology, disciplines that were unknown to Enlightenment thinkers. Intuitively, it makes sense that humans, as a species that evolved in small groups, would have some degree of altruism. The successful individual is one who can take into account the well-being of others in the group. Natural selection would favor that trait.

In fact, we see this in ordinary life as well as in laboratory studies. Babies quickly learn to share.

When a baby is upset, his playmates comfort him.

Grasp of goals

In a study conducted by Felix Warneken and Michael Tomasello, a toddler watched a man he knew trying to put an object into a closed closet (Felix Warneken and Michael Tomasello, "Varieties of altruism in children and chimpanzees," *Trends in Cognitive Sciences*, Vol. 13, No. 9, pp. 397-402). The man pretended to be baffled and continued to bump into the closet door. The toddler, who could walk but not yet talk, wobbled over to the closet and opened the door for the man. While lacking language, the toddler had a grasp of goals. He had at least some understanding of what the man intended to do.

Bloom argues that this study and many others show that babies have a rich understanding of the physical world and social world.

Investigators are asking now whether they have a moral understanding — not whether they are inclined to do good, but whether they can tell good from bad.

Bloom says the one-act plays with puppets showed that the children react strongly, even before they acquire language, to good and bad behavior. Children have, in his view, the capacity to make moral judgments before they acquire language and the ability to state moral reasons.

The missing cat

Bloom offers an insight from a case of a man in London who was feeding a cat. The cat came every day for his meal. When the cat didn't show one day, the man became concerned. He searched everywhere and finally, when he was taking out the trash, discovered the cat in the garbage bin. How did the cat get into a bin with a lid? Surveillance cameras are common in London, and when the inevitable recording was found, it showed that a woman came by, saw the cat and put it in the bin.

The man posted the video on social media, and the woman was tracked down and shamed. Journalists appeared at her door. In her defense, the woman uttered the unfortunate phrase, "It was just a cat." The public reaction was so strong that the woman eventually had to have police protection. Bloom cites the episode as evidence of a human need to reward, publish and punish. Adults, armed with social media, are much like the toddler who slugs the naughty puppet who won't share. We want to punish the bad guy and reward the good.

Friend *v.* stranger

There's a darker side to this impulse. Evolution apparently inclines us to favor family and friend over the stranger. It's an adaption favored by our evolutionary history in small societies. Fear of the stranger was a useful trait in an era when it was often suicide to wander away from one's own band. Babies don't welcome strangers. When a baby is dropped into Santa's lap, the baby usually cries.

This phenomenon is known as "stranger anxiety." From an evolutionary perspective, it's easy to see the advantage to the group that hordes resources and is suspicious of strangers. A parent who favors his or her own children gives them an advantage when compared to a parent who loves all children equally. Similarly, small-scale societies treat strangers as competitors for resources. Margaret Mead, in describing her work among small-scale societies, said: "Most primitive tribes feel that if you run across one of these subhumans from a rival group in the forest, the most appropriate thing to do is bludgeon him to death."

Babies have stranger anxiety, and the story gets worse in the later stages of childhood.

Moral thought and sharing

One interesting study that highlights these attitudes in older children involves a game with poker chips, which can be redeemed for prizes. The simplest version of the game involves a 1:1 ratio — one chip for the child, and one chip for another, unknown child.

The game involves the child making choices about apportioning chips between himself and the stranger.

So, as the game evolves, does the

child choose a 1:1 ratio or a 2:2? That's easy. It's self-interest. Two chips are better than one. But what about a 1:1 distribution versus a 2:3 ratio? Wouldn't the child choose to get twice the number of chips for himself, even it if meant giving even more to the stranger?

Up until the age of 8 or 10, children choose the lower number. They are willing to sacrifice real resources to maintain a relative advantage against a stranger.

Where does that leave Adam Smith's claim about human nature including a concern for the interests of others?

The tendency toward helping and sharing is true — when the other people are familiar. However, when other people are strangers, it's a darker picture.

Emerging capacity to reason

Bloom's view of moral capacity is that we have one, but it's limited.

Our moral capacity is limited in ways that are similar to those of other primates.

But that summary also comes with a puzzle: We're better than that, and the evidence of that is not hard to find.

In the state of nature, we humans care only about our families. But we're not limited to that. We come to understand that morality extends to everyone on earth, and perhaps to other living creatures.

While there obviously are large cultural differences, philosopher Peter Singer has pointed out that awareness of the morality of our treatment of animals is almost universal. Healthy humans just don't torture cats. We also seem to be evolving as a species toward less violence, not more. As the cognitive psychologist Steven Pinker has pointed out, the evidence is strong that human violence is declining (Steven

Pinker, *The Better Angels of Our Nature: Why Violence Has Declined*, Penguin Books, 2012).

Rational capacity

Bloom says the more mature version of morality emerges with our rational capacity. As babies, we have sentiments and reactions. But as we develop rational faculties, we use the power of rationality to curb our sentiments and emotions.

As Adam Smith put it, "it is not the soft power of humanity, it is not that feeble spark of benevolence which Nature has lifted up in the human heart. Rather, it is reason, principle, conscience … that calls to us, with a voice capable of astonishing the most presumptuous of our passions, that we are but one of the multitude, in no respect better than any other in it."

This is known as the principle of impartiality, the realization that we are no better than others. That principle shows up in every religion, every moral philosophy.

Each of us loves his or her own children more than the neighbors'. But as a moral matter, each of us realizes that our own children have no more moral weight than any others. We support impartiality — fair and equal treatment — when our own children go to school.

Balancing interests

Rationality, needless to say, is a powerful tool. It allows us to put our own petty interests aside to see a larger good. That, Bloom says, is the dawning of morality.

In his book *Against Empathy,* Bloom explores this duality in our moral life: We are capable of reason, but we are pulled by biases.

Most of our biases are things we'd want to strongly disavow. Racist or xenophobic tendencies are prime examples. Most of us would deny that we make decisions based on the attractiveness of another person, though studies suggest we are, in fact, influenced by another person's appearance.

Bloom said the sentiments that seem to be most precious, like empathy, can lead us astray. We feel so strongly about our own children that we tend to think they are most qualified for any job.

But our rationality allows us to see the problem with those sentiments and to devise methods to curb them. We devise nepotism rules because we recognize the dangers in our own nature. We develop constitutional governments for the same reasons.

Decide or justify?

Many scholars are dubious of the role of rationality in making moral decisions. Jonathan Haidt argues in his book *The Righteous Mind* that we use our capacity to reason to justify moral decisions we've already made (Jonathan Haidt, *The Righteous Mind: Why Good People Are Divided by Politics and Religion*, New York: Vintage, 2013). Haidt says we see ourselves as impartial judges, but we're really more like lawyers: we get our case and argue it.

Bloom argues for a stronger role for reason in our decisions. We know we have biases. We use our intelligence — our rational capacity — to override them. We are smart enough to create institutions to get around our individual limitations.

"I think people often get two things wrong about morality," Bloom said. "First, many people believe that we are born selfish and amoral — that we start off as natural-born psychopaths. As you can see, there is plenty of evidence that this is mistaken.

"Second, many people believe that we are, as David Hume put it, slaves of the passions. Intelligence and wisdom are largely impotent; we're just driven by our emotions. Now, this is a very tempting view of human nature given the nature of this election season. And if it's true, we should buck up and learn to live with it.

"But I argue that it's not true; these dismissive claims are refuted by everyday experience, by history, and by the science of developmental psychology. We are moral animals, but we are also *rational* animals, and we are powerfully influenced by our capacity for reason."

Reasoning faculties

When and how does that rational capacity develop?

In humans, the basic architecture

of the brain is in place after the first few years of life. That means that if you look at a 6-year-old's brain with the naked eye, it wouldn't look all that different from an adult's brain, said Silvia A. Bunge, a neuroscientist at the University of California at Berkeley. Bunge has studied how the brain changes as the ability to reason emerges between the ages of 6 and 20.

While the brains of the children and adolescents appear similar, measurements with MRI technology tell a subtler story. "You will find that the connections between the brain regions that are involved in reasoning get stronger over childhood and adolescence," Bunge said.

Bunge joined the lab of John Gabrieli as a graduate student in 1997, in the early days of functional MRI. Up to that point, most of the studies using this technique had been done in adults. She was interested in studying how the brain works and realized that studying how the brain matures would provide insights into how it functions in adulthood. "And then I gradually discovered how fascinating brain development is in its own right — and how important it is to understand it," she said.

Bunge has a particular interest in the capacity for reasoning, noting that it is essential for learning, and learning is essential for reasoning. Indeed, studies show that reasoning ability in childhood is linked to achievement in higher education and in demanding careers. "We are beginning to understand how the brain reasons, why we get better at it over childhood, and how our minds and brains change when we practice our reasoning skills," Bunge says.

"Every one of us has to reason our way through problems in our lives, as well as in school and at our jobs. However, people vary widely in their reasoning skills, and so it's important to understand whether, when, and how we can improve at it. This line of research is relevant for people who are curious as to how our brains change, and how these changes support cognitive development."

Exercising the mind

Bunge's work suggests that exercising the mind can develop the brain and make a person a better student, just as lifting weights and running sprints can develop the body and make a person a better athlete.

"Some researchers don't like the analogy between exercising the mind and exercising the body, but I think it's quite apt," she said. "In both cases, prolonged and repeated training is necessary. Benefits are hard-earned, and fade after you stop 'working out.' And the benefits are quite specific, just like with weight-training: you can't lift weights with your arms and expect your calf muscles to get stronger. Likewise, you can't practice one kind of cognitive skill and expect to see benefits across the board. Another reason I like the analogy is that … athletic training has improved as understanding of the body has accumulated. We're still at the early stages of figuring out how best to boost learning and brain plasticity. Researchers are experimenting with various approaches, for example combining cognitive training with aerobic exercise training, or with transcranial magnetic stimulation, or even with pharmacology, and there have been some promising early results."

One obvious way of giving the mind some exercise is by reading. Bunge said she's concerned that many teenagers don't read books for pleasure. "Reading requires sustained attention (sitting still and focusing on the printed word when there are so many distractions), working memory (keeping track of who's talking or what's happening from one page to the next), and long-term memory (remembering the plot line when you pick up the book again after some time away), and it provides the opportunity to build vocabulary," she said. There are other benefits, she said, citing former President Obama's observation that he liked to read novels because they had helped to develop empathy.

"It is no exaggeration to say that every experience leads to changes in connections between neurons in our brains," Bunge said. "There's a great quote from Will Durant that's been falsely attributed to Aristotle: 'We are what we repeatedly do.' Excellence, then, is not an act, but a habit.'"

The adolescent brain

The story of the development of the human brain is not just about the emergence of a capacity to reason. As human beings mature, they develop a capacity for well-being. They develop a capacity to think about happiness — as well as to be happy.

What does it mean to live a good life and be happy? The pursuit of that question requires adolescents to use their reasoning capacities and their moral sense. It also requires the adolescent, who has grown up as part of a family, to find his or her own way. The question is, in a way, the route to maturity as an adult.

Robert Duke and Arthur Markman, professors at the University of Texas and hosts of "Two Guys on Your Head," an

NPR radio show that examines some of the baffling aspects of human behaviors, describe happiness as a mix.

You can attribute about half the differences we see in people in terms of their happiness to genetic factors. About 20 percent is a matter of what's going on in a person's life recently — we all have good and bad days. The remaining 30 percent of the happiness formula involves things that we largely have control over — things that we can do that can move us, long-term, in the direction of happiness.

Problems of interpretation

One of the difficulties of studying happiness is that each of us has a different interpretation of what happiness is. People who are extroverted, buoyant and loud have a view of happiness that quiet, introverted people just don't share. We tend to assume that the definition based on our own personalities should fit all.

Leo Tolstoy, in his great novel *Anna Karenina,* posited that "All happy families are alike; each unhappy family is unhappy in its own way." In Duke's and Markman's view, Tolstoy was wrong. Not every happy person is happy in the same way. Happiness is subjective. It depends upon the person having the experience. When it comes to happiness, each individual tends to construct his own brand.

Constructing happiness

One of the themes that Duke and Markman emphasize is that we humans have a remarkable capacity to construe things. Some people find the tiny dark cloud spoiling the enormous silver lining, while others find the invaluable silver lining in the darkest, most threatening cloud.

Duke and Markman suggest that a good place to start in a personal exploration of happiness is to ask: What do you tend to find?

When we are in a positive mood, we tend to find positive things. When the sun is out and we're in the company of friends, everything seems to go well. What that suggests is that moods have a propensity to perpetuate themselves. That, in turn, suggests that if you want to lift your mood, just make it a habit to look for positive things.

That's the first of several strategies for cultivating happiness. Cultivate the habit of looking for — and appreciating — positive things.

Ditching an illusion

A second strategy involves getting rid of an illusion. Almost all of us come to the intuitive view that our minds are in our brains, which are in our skulls, which are part of our bodies.

In fact, our minds are not separate. Our minds emerge from and are shaped by nerves that tell us we have a thorn in our toe and by complex systems that tell us we are hungry and that a neighbor is grilling hamburgers. Our minds also are shaped by our environment, particularly by friends and loved ones.

And, since our minds are not separate from our bodies, our physical well-being is not separate from our emotional well-being.

To take one example, research suggests that Americans might be among the most sleep-deprived populations in history. We tell ourselves we can work on four hours of sleep a night, when in fact we can't. We can't function emotionally or intellectually if we're not sleeping.

It takes a conscious effort to change habits that are undermining well-being and, thus, happiness. But the good news is that by taking some relatively simple steps — such as getting adequate sleep — you can, in some cases, heal yourself.

Genuine effects

Any person who investigates his or her own happiness gets to this question: What is *genuinely* affecting my mood?

The fear is that, if I accentuate the positive, I'm not being authentic. If I try to look on the bright side of a tragedy, I'm just sticking my head in the sand.

Duke and Markman acknowledge that's true. But they also point out that while tragedies happen, they generally don't happen several times a day, every day.

Moreover, this concern about

authenticity is sometimes based on a false premise about the brain. It's simply not the case that everything emanates from the brain outward. If you're genuinely baffled by what's contributing to your mood, work from the outside in, rather than inside out. For example, there is a small physiological connection between your mood and smiling. The physical act of smiling lifts your mood a bit.

This act reinforces itself in a secondary way. Other people are programmed to mirror your mood. It's human nature to rejoice with those who rejoice. From birth, humans imitate. If you walk around with a frown, you prompt others to frown. Smile, and most of the world will smile with you.

Baseline happiness

We all know people who seem to be happy naturally and people who seem to be naturally less happy. In summarizing the research on happiness, Duke and Markman said we all tend to return to our individual baseline.

There is no overall baseline for happiness. And that means that each person must come to terms with his or her own.

The idea that each of us tends to return to his own baseline of happiness has implications.

First, it suggests you shouldn't be afraid to put yourself in situations where you might fail. We often talk ourselves out of facing challenges by telling ourselves that we'll feel bad if things don't work out.

Duke and Markman cite research by Daniel T. Gilbert and Timothy D. Wilson, who studied faculty members who were being evaluated for tenure. The researchers talked to candidates for tenure before and after they were reviewed.

Six months before the review, the candidates were asked how happy they would be if they were granted tenure. The usual reply was that they would be really, really happy. If they were not granted tenure, they would be profoundly unhappy. For academics, tenure is generally viewed as a make or break deal.

But, after the review, that didn't turn out to be the case. Some of the candidates got tenure. Some did not. But, on average, everyone was equally happy.

A day after the tenure decisions were announced, some were joyous and some were devastated. The difference was large. But six months later, that difference was not anywhere near so large.

Perspective

If you look at happiness in terms of human development, an obvious fact leaps out: Younger people don't have that long-view. In children, time scales are short. Joy is immediate and wonderful. Five minutes later, it's possible to have a tragedy.

Among the wonders of the human brain is that it puts a time machine in our heads. We can go into the future through imagination and into the past through memory. We can remember great tragedies and great moments of joy. Over time, those extremes even out.

That gaining, or maturing, of perspective is a natural process. As we get older, we don't see every little thing as a wonderful triumph or a devastating loss. The circumstances of our lives don't generally change in 24 hours. There *are* exceptions. Real tragedies do happen. But they are not everyday occurrences.

In general, our ability to handle

the ups and downs of normal life is a picture of our mental health. As we get older, our emotions stabilize. Especially in developed countries, our lives take on stability and predictability as we mature.

'Hedonic treadmill'

Even as our lives become more stable with maturity, we are not good at predicting what will happen in terms of improving our moods. In a consumer society, we are bombarded with suggestions that buying a new car or TV will make us happy.

In the moment you buy the new toy, you are happy. But then you find yourself on the "hedonic treadmill," a concept proposed by Philip Brickman and Donald T. Campbell in 1971. (Brickman, P., & Campbell, D. T., "Hedonic relativism and planning the good society," in M. H. Appley (Ed.), *Adaptation-level theory* (pp. 287-305).

New York: Academic Press, 1971).

As soon as you get the new thing, it's part of your stuff. Your life adjusts to having that as the routine, so you must find something else to want, some new toy to acquire. Buying stuff does not make you happy in the long run.

That said, money well spent can make you marginally happier in at least a couple of limited ways.

First, you can pay people to do things you really don't want to do. Those with allergies might find it liberating not to have to mow the lawn.

Second, when spending your money with happiness in mind, buy experiences, rather than stuff. The brain does give humans that remarkable ability to time travel. You can create memories by putting your resources into special occasions, vacations and journeys to interesting places.

Other people

Duke and Markman point out that at least some of your happiness depends on the choices you make. The people you choose to surround yourself with do affect your happiness. People who share your values are much more likely to contribute to your happiness than those who don't.

In addition, some close friends and loved ones can help you manage your moods when you are not doing well at it. Duke says that his wife plays that role for him. Duke says he loves being a professor and researcher but hates committee work. After he's endured a committee meeting, his wife will give him a minute to vent and complain. But only a minute.

Most people think that being an empathetic person involves listening to the complaints at length. But an hour-long wallow in frustration doesn't mend your mood. Being that kind of "sympathetic" listener just enables a person to remain in an unhappy state. There's a point at which nothing more can be said that would make a situation better. It's good to have people who know you and can help you get out of the emotional muck.

Strategies for sadness

Sometimes, feeling bad feels good — at least a little bit and at least in the sense that you don't want to make the feeling go away. For example, remembering a loved one who has died is sad, but we might want to savor the memory.

Our moods are part of our evolutionary history. Our moods let us know how things are going. Our moods derive from connections between the frontal lobe and deeper parts of the brain. One part of the brain can't tell the other what to do.

When you're sad and someone tells you to "Snap out of it," it's just not helpful. Your cortex can't tell your basal ganglia what to do.

You have to *experience* your way out of it.

We can, to an extent, control our environment. We can choose what kind of music we listen to. We can choose what kind of people we hang around. We can choose whether we spend the morning in the sunlight in the park or in the dark, damp basement. We can't order our brain to "snap out of it." But the experiences we choose for ourselves *can* get to the more inaccessible parts of the brain.

Practicing happiness

The question of human happiness is so complex that some people want a shortcut. That usually means they want to find a drug, such as dopamine or serotonin, that might make them happy.

Duke and Markman argue that happiness has more to do with our interaction with the world than the possibility of finding a magic drug.

One class of drugs commonly prescribed for depression is known as selective serotonin uptake inhibitors. These drugs work by increasing the influence of serotonin in the brain. When you take a pill, nothing happens immediately. Over two weeks, you begin to notice your mood is changing.

If happiness were just a matter of one chemical, it shouldn't take two weeks to work. If you had a runny nose, you wouldn't take a Benadryl if it took two weeks to work.

Instead of seeking a magic switch to flip, Duke and Markman argue for long-term, repetitive practices. Both, as musicians, are used to practicing to improve their music.

Both argue that all of us, as human beings, can cultivate practices that are better ways of responding to the world.

The things we practice — the things we do repeatedly — tend to become the way we respond to the world. For example, as we get older, most of us learn that it's never a good idea to send an angry email after dinner. You can allow yourself to write the email. It feels good. But then you know not to send it. After a night's sleep, you still don't usually send the email. And, if you do, you edit it. That's one example of one small practice that can make life happier. You break an old habit and teach yourself a new one that can influence your circumstances for the better.

Too often, we just respond. We don't realize we can *choose* how we respond.

Adolescent challenge

All human beings pursue happiness, but the challenge seems especially daunting for adolescents. What's going on in the adolescent brain?

Duke and Markman suggest three things.

First, mood is a reflection of the human motivational system, and the motivational system, which often is not a model of clarity in adults, is confused in adolescents. An underlying structure of the brain, the basal ganglia, is part of the brain's motivational system. Circuits in the basal ganglia are not well connected to those in the cortex, the section of the brain associated with reasoning, and Duke and Markman point out we are often not well connected to our own motivations. Sometimes, we need therapeutic help to find perspective on why we do what we do. The motivational system works through feelings. When things are going good, you feel good. When not, you feel bad. Your feelings are an index of your motivational system.

A lot of research has focused on "life tasks." The idea is that at different points in your life you have different tasks. In your early 20s, your focus may be on starting a career or a family. But, as an adolescent, your main task is to establish your own identity, to become a unique human being rather than just an appendage of your family.

It's difficult and confusing when your primary task involves a paradox. Your primary life task is separating yourself from others, but you know that human connections — supportive connections to others — are essential to well-being. That paradox is one of the reasons adolescents are not the happiest people on earth. Nevertheless, the life task of the adolescent is to pull away from one support structure to become an independent adult.

A second complication with adolescents is that there's a tremendous amount of physiological change, and our evolutionary history is not in line with our current cultural standards. Just as adolescents are negotiating this paradox of breaking away from family to become individuals, they experience an influx of hormones that set goals adolescents can't really act on. When you are an adolescent, your body is telling you to have sex, to procreate to ensure the survival of the species. But social controls are shouting no.

Keep in mind that the human brain evolved in different times. In the early days of human evolution, life spans were short, and humans, like all living things, evolved to reproduce the species. Throughout the vast majority of human history, people reproduced at an early age.

Now, things are different. When an adolescent shows the self-restraint society demands, he or she is blocking a goal set by the motivational system. When the motivational system is frustrated, moods are bad.

At a time when adolescents must separate themselves from their family support systems and become independent people, their physiology is wreaking havoc with their motivational systems. The motivational system is experienced through feelings — and it is felt deeply.

In addition, adolescents have a limited perspective on time. An adolescent's life has, by definition, been shorter than the adult's. An adolescent may see today's top event as the greatest thing that's ever happened. The event that occurs tomorrow may result in the worst day ever. The adolescent brain is forced to ride a rollercoaster. Adolescents experience intense feelings, which are in conflict because cultural expectations conflict with evolutionary biology, and they lack the experience that would provide them with a longer viewpoint.

What happens between the periods of development and decline? The most remarkable feature of the adult brain is the mind that emerges, a mind that continues to learn. We'll talk more about that in Part II.

Chapter 2

Cognitive enhancement

Scientific advances have improved our understanding of the brain. Can these same advances improve the brain's performance?

It's no secret that stimulants can enhance cognitive performance. If you doubt that, ask some kids in junior high and high schools. A study published in 2008 suggested that about 25 percent of U.S. students in secondary schools were using psycho-stimulants. And that's just counting the kids who have *no* psychiatric or neurological disease.

Is that wrong? In cases where there is no disorder, no disease, is it wrong for a physician to prescribe a drug? Or, if you're a teenager who doesn't have a prescription but has connections, is it wrong to take such a drug?

Walter Glannon, a philosopher at the University of Calgary and the author of several books on the ethical implications of advances in neuroscience, says the topic of cognitive enhancement is perhaps the most hotly debated area of bioethics today.

The idea of "enhancement" goes beyond therapy. It's an attempt to optimize functioning — taking what is normal and turning it into something that is even better.

We've been dazzled by the advances in science. But before we go further, we ought to ask ethical, rather than scientific, questions: Is cognitive enhancement fair, right or good?

Three scenarios

Glannon suggests it's helpful to consider three scenarios.

The first involves a student who takes Ritalin, a drug for Attention Deficit Hyperactivity Disorder, to help him concentrate on exams. The student does not suffer from ADHD, but the drug works. He gets better grades than the other students in his class. Is that fair? Or is it cheating?

The second scenario involves an airline pilot who takes Provigil, a psycho-stimulant, to remain alert during long flights. Obviously, the thought of a pilot dozing off during a long flight is disturbing because it puts so many people at risk. Isn't it better — almost a public good — for pilots assigned to those flights to take such a drug?

The third scenario involves surgeons who are working longer hours and performing more surgeries. Some take Provigil and Adderall to stay awake and alert. If you are a patient and are scheduled for surgery late in the day, can you express a preference that the surgeon take such a drug?

Intuitively, the first case strikes many people as simply wrong — a case of cheating. The student, in taking a drug, is relying on something other than his own effort to achieve his goal and in doing so is getting a competitive advantage over his peers. The case is similar to doping scandals in sports.

But the same intuitions don't seem to apply to the airline pilot and the surgeons.

Why is that? Why do we make a distinction in one case and not the others, when in all the cases, people are doing something other than using their own natural capacities to enhance their cognitive functioning?

Developing guidelines

Glannon says that we have a complex web of social norms and expectations about what we do to ourselves. And while we've reached a point where a significant number of secondary-school students are using drugs to enhance performance, we haven't thought through these advances in science that have individual and collective consequences.

In an article published in 2008 in *Nature,* several lawyers and policy experts offered some general guidelines for developing laws on cognitive enhancement. One obvious suggestion was to provide legal protection against coercion. It would be one thing if a surgeon *wanted* to take a drug to enhance performance. It would be another if a hospital *coerced* a surgeon to take a drug so that she could perform more surgeries every day to generate revenue. But in the decade after the article was published, Glannon has seen little follow-up, despite evidence that more drugs are being used off-label to enhance cognitive performance.

What would good policy and good law look like?

So little is known about the long-range effects of psycho-stimulants, the obvious place to start would be to point out the importance of commissioning such studies. More good scientific information is needed.

But, with so many people using drugs that enhance cognition, it's time to start public discussions about public policy. For example, could the widespread use of such drugs be a public good? One pattern that has emerged in the research about the effects of psycho-stimulants is that they seem to have a leveling effect.

Those with lower cognitive functioning tend to benefit more. Those with higher cognitive functioning tend to benefit less. If further research confirms that pattern, is that ethically significant? Would the widespread use of such drugs level the playing field and make the world a little bit fairer?

And if that idea is abhorrent to you, what is the rationale for a ban on such enhancers?

Rules and marketplaces

A couple of decades ago, cognitive enhancement was on the horizon, and people were just beginning to speculate about what it would mean ethically and legally. Today, cognitive enhancement is here. Chances are, it's commonplace in your high school. Moreover, a pattern has developed in which new drugs that show promise in improving cognitive performance somehow get into circulation even among people who have no diagnosed diseases or disorders. A German psychiatrist found that 89 percent of the patients who had a prescription for Modafinil (Provigil) did not have a psychiatric or neurologic disease (Franke, Andreas G.; Dietz, Pavel; Ranft, Karla; Balló, Harald; Simon, Perikles; Lieb, Klaus, "The Use of Pharmacologic Cognitive Enhancers in Competitive Chess," *Epidemiology*, November 2017, Vol. 28, Issue 6).

The patients were using the drugs simply to enhance normal function.

In early 2018, scientists announced they'd developed a brain implant, which functioned like a pacemaker in that it sent electrical impulses to aid the brain when it sensed trouble with normal functioning. At the time, researchers were discussing commercial possibilities for the technology. They also cautioned that the possibilities for misusing the technology were vast. If the technology is approved for patients with dementia and brain injury, what about people who simply want better cognitive performance? Won't there be pressure to make such technology available through a free marketplace?

Self-regulation

Glannon has suggested that the reluctance to regulate drugs and therapies that enhance cognition could be problematic for the same reason that it's problematic for a patient to decide how much morphine he or she needs.

Deep Brain Stimulation — a technique that involves implanting electrodes and stimulating areas of the brain with electrical current —

has proven so addictive in laboratory mice that some will literally stop eating. The procedure raises dopamine concentrations to addictive levels. Given a choice between Deep Brain Stimulation and food, mice choose stimulation.

Glannon cites the case of a patient with severe depression who had a course of Deep Brain Stimulation. The symptoms were relieved, and the patient told his physician that he felt good and wanted to feel even better. He asked for an increase in the frequency. The patient became euphoric but reported that he felt unrealistically good. He asked that the frequency be stepped back. As Glannon put it, this patient caught himself before he lost control. But the case history illustrates the risks.

Knowledge of risks

Even people who are interested in much milder therapies and drugs may simply lack the information to make good choices. Some drugs and therapies have tradeoffs. A drug that may improve your ability to process information may inhibit your ability to make decisions. A study using Transcranial Magnetic Stimulation, a technique in which the brain is stimulated by electromagnetic waves, showed that certain therapies enhanced the ability to *learn* mathematical lessons but inhibited the ability to *apply* them. At the very least, the new technologies provide some new challenges in informing consumers.

Finally, we know little about the long-term effects of altering function in developing brains. Some parts of the brain, including the frontal lobes, are not completely developed in adolescents, who seem to be the most aggressive experimenters with cognitive enhancing drugs. Younger people tend to be less averse to risk. It's an open question about whether juveniles, at least, need protection.

The public has been slow to grasp some of the implications of these scientific advances. What's needed is a way of getting information about new developments in neuroscience to a broader audience and a broad discussion about the best way to harness the new drugs and new therapies as a public good.

Chapter 3

Mapping the brain

In 2016, our understanding of the brain took a revolutionary step forward. A map published by the Human Connectome Project detailed 180 regions of the brain, 97 of which hadn't been known.

David Van Essen, a Harvard-trained neuroscientist who has described himself as "a cortical cartographer," led the Human Connectome Project, based at Washington University in St. Louis.

He began his work in the 1960s,

using a pencil and tracing paper to try to map the areas of the brain that were then known. He's followed that interest through revolutionary changes in imaging technology and computing. When he began, it simply wasn't possible to map a living brain. Now, with functional MRI technology, it's possible to get images of people doing different things: listening to stories, doing arithmetic, looking at images and kissing. We can, in other words, see what regions of the brain are involved with different actions.

Many different kinds of maps are possible, including maps that go beyond physical structure to show various functions.

"A useful analogy is to think of the maps of the Earth," Van Essen said. "Some show mountains and rivers and other geographical features. Others show countries, states and other political subdivisions. Each serves a different purpose. When we make maps of the folded cerebral cortex, these are analogous to geographical earth maps. Our maps of brain function are akin to maps of political subdivisions.

"Early Earth maps were quaint and amusingly inaccurate, but now we enjoy the extraordinary precision of Google Maps. Brain maps have dramatically improved as well, though not to the same degree. In both domains, map-makers share similar fundamental objectives, but the tools continue to evolve."

Are there limits?

The word "connectome" —

which refers to the set of connections in our brain — is a reference to "genome," which is the sequence of 3 billion base pairs that make up 21,000 genes. Collectively, they're the instruction kit for making our brains and bodies.

While scientists such as Francis Crick expressed confidence that we'll come to understand consciousness, just as we have come to understand the human genome, Van Essen has expressed doubt.

"We have made great progress in characterizing the human genome. This includes incredibly accurate mapping, or sequencing, of all three billion 'base pairs,' the genome's alphabet, in the genomes of each of many thousands of people. Yet there are still vast mysteries about how genes are turned on and off to orchestrate the development of healthy cells, organs, and people or what goes wrong in most genetic disorders.

"In the brain, there are 80 billion nerve cells and trillions of connections, or synapses. Accurately mapping all those cells and their connections — what we call the connectome — is currently impossible for the human brain, though it may become feasible for laboratory animals such as fruit flies and even mice. Nonetheless, there are countless exciting discoveries about the human brain still on the horizon."

Among the questions on the horizon are how and why the cortex of brains in more advanced organisms, such as mammals, fold in characteristic patterns. Those patterns are specific to species. While some of the patterns are stable, many vary, even in identical twins. It's a puzzle: two individuals with the same genetic instructions have different patterns of folding in a key structure of the brain.

Basic research

In 1909, German anatomist Korbinian Brodmann, using stained slides, published drawings of about 50 cortical areas in humans. With better tools, we are getting a better picture of how our brains work to make us unique — as a species and as individuals. We know things about the brain and its 86 billion neurons that Brodmann could not have fathomed.

Van Essen said there are two reasons to pursue basic research on the brain.

"The first is human curiosity. How do our brains work? What makes us think, sing, and feel love or hate? Such questions fascinate the lay public as well as neuroscientists.

"Second, we must pursue this basic research in order to better diagnose, treat, and prevent brain disorders and diseases. All of us have had the experience of seeing someone with a debilitating brain disorder. We need to sustain our efforts to identify cures and provide better treatments."

Finding your sustaining passion

In discussing the future of neuroscience, Van Essen talked not just about developing technology but about developing the next generation of scientists.

Van Essen found his field of study when he came across a book on the brain after his sophomore year at California Institute of Technology. He urges students to keep exploring until they find a field that sparks their passion.

"I was interested in science from early childhood. In high school, I was fascinated by math and physics. In college, I realized my brain wasn't really 'wired' to become a theoretical physicist. I became a chemistry major instead, but chemistry didn't ignite the kind of passion that sustains a career. By good fortune, I chanced across a book about the brain and immediately became hooked.

"One lesson is that success in science requires intense work but also a fierce passion for your area of study."

Chapter 4

Neuroplasticity

A few decades ago, biology students were taught a dogma that has proved to be fundamentally wrong. The human brain was conceived as the end product of a long evolutionary tree. It was complex but fragile. The human brain had the capacity to do things that were far beyond the capacities of the brains of other primates, our nearest evolutionary relatives. But the price of that wonderful sophistication was fragility. If anything broke inside the brain, it was impossible to repair. The brain grew and developed, but when it was mature its anatomy was set. The brain did not really have the capacity to change its structure or to heal itself.

That was the prevailing view when many of today's leading scientists were in college. Since then, scientists and physicians such as Dr. Nicholas D. Schiff and Dr. Joseph J. Fins have documented case after case about how even badly injured brains sometimes heal. The concept of neuroplasticity, the idea that the brain and nervous system can change, went from being ridiculed to being accepted.

Dr. Norman Doidge, an internationally renowned psychiatrist and psychoanalyst, has written two best-selling books on neurosplascity: *The Brain that Changes Itself,* which sold more than a million copies, and *The Brain's Way of Healing.*

Doidge says the old idea that the brain's complexity made it fragile, fixed and unchanging was simply wrong. He also says the idea that the brain is permanently hardwired like a computer is similarly wrong.

His books include case histories of people whose conditions were thought to be hopeless. One case involved a woman who was born with half a brain, which reorganized itself to function as a whole. Other cases involved people with strokes, cerebral palsy and low-learning disorders.

Tastes and loves

Neuroplasticity is an enormous

concept. Its possibilities in thinking about healing are intriguing. But consider what it might mean as we think about what the human brain does as a matter of routine — and also what it might be capable of.

One of the most interesting features of Doidge's work is his account of how we acquire tastes and loves. It's an account of how we become interested in other people and in other things.

He begins with a basic instinctual interest: sex. All animals have a drive to reproduce, but humans have an extraordinary degree of sexual plasticity compared to other species. The range of what people find pleasurable and desirable is remarkable. Doidge says that Sigmund Freud, a pioneer in psychiatry, was right that an individual's tastes in sexuality and intimacy begin in early childhood, rather than at puberty. "If the parent is warm, gentle, and reliable, the child will frequently develop a taste for that kind of relationship later on; if the parent is disengaged, cool, distant, self-involved, angry, ambivalent, or erratic, the child may seek out an adult mate who has similar tendencies" (*The Brain That Changes Itself*, 98).

In other words, the idea of what is sexually desirable is not hardwired into the brain. It's influenced by our experiences.

That fact implies that what is wired into the brain can, over time, be changed. Doidge cited the case of a man who pursued destructive, aggressive women because his childhood experiences had fused the experiences of sex and aggression. Neuroplasticity, the malleable quality of his brain, had allowed that fusion to occur. Neuroplasticity allowed that idea to be unlearned and a better idea substituted in its place.

The notion of "love" is similarly flexible. Romantic love, now the norm in the West, is not universal and was a distinctively odd idea until recent times. The notion of the right to choose one's own spouse became the norm only after democratic ideas and institutions were widely adopted. Now, the idea that a person has the right to choose his or her own mate is considered "natural."

That adaption is possible because of the remarkable plasticity of the brain. As Doidge puts it: "The brain structure that regulates instinctive behaviors, including sex, called the hypothalamus, is plastic, as is the amygdala, the structure that processes emotion and anxiety. While some parts of the brain, such as the cortex, may have more plastic potential because there are more neurons and connections to be altered, even noncortical areas display plasticity. It is a property of all brain tissue" (ibid., 97).

Learning and unlearning

To understand plasticity, it's important to grasp that the process of unlearning is as important as the process of learning. Intuitively, it makes sense that you would have to get rid of some unhelpful ideas before you could absorb ideas that are more useful. In fact, different chemical processes are involved in learning and unlearning.

The learning process, described at the neuronal level, is called long-term potentiation, or LTP. As Doidge puts it, neurons that fire together wire together. Neurons make connections through repeated firings. That's how we learn.

The process of unlearning involves disconnecting those channels. It's called long-term depression, or LTD.

Walter J. Freeman, a professor of neuroscience at Berkeley, points to massive neural reorganization at two points in life: falling in love and parenting. That reorganization involves unlearning old ideas and behaviors, as well as learning new ones.

The process is aided by neuromodulaters such as oxytocin,

the so-called "commitment" hormone. It reinforces bonding in mammals. Oxytocin is found in high concentrations in new mothers, and the closely related vasopressin is released in new fathers.

While the neurotransmitter dopamine is all about excitement, oxytocin is associated with calm, warm, nurturing moods. In evolutionary terms, this is a remarkable development. Nature provided a mechanism that helps two brains to work closely. Each partner reads and shapes the other's perceptions and intentions.

Rules of plasticity

In Doidge's account, a basic rule of plasticity is that if one brain system changes, the other systems that are connected to it change.

When a young man becomes a father, a massive reorganization goes on in the brain. The young man unlearns old ways and learns the new realities of caring for a child.

Another basic rule of plasticity is that neurons that fire together wire together. That's essential because we humans use more than one area of the brain even when doing relatively simple tasks. That wiring allows different areas of the brain to work together.

One of the implications of that rule is perhaps the most basic rule of plasticity: Use it or lose it. If these neural connections are used often, they become stable. If not, they may be "unlearned" or disconnected.

Pleasure and pain

One other aspect of human nature is needed to complete the picture of how human beings acquire tastes and love for new people and new things. About 70 years ago, researchers reported on "pleasure centers" in the limbic system, a part of the brain involved in processing emotion. Dr. Robert Heath implanted electrodes in the septal region. The electrical stimulation created euphoria in patients.

Heath found that when pleasure centers are firing, it's more difficult for a person to experience pain or displeasure. That's why, when we're "in love," the feeling is close to intoxication. We talk all night. We have endless energy. We also may have impaired judgment. We experience so much pleasure that our brains give us less than a full, clear picture of our beloved's flaws and bad behaviors— things we normally would be averse to. "Being in love" is cool in itself, but it also makes it harder to be unhappy.

As Doidge says, the image of the new person "becomes associated with unexpected pleasures and is plastically wired into the brain, which has evolved to respond to novelty. We must be learning if we are to feel fully alive, and when life, or love, becomes too predictable and it seems like there is little left to learn, we become restless — a protest, perhaps, of the plastic brain when it can no longer perform its essential task." (ibid., 116)

Interests

What is true for matters of tastes and of loves is also true for matters of interest.

Just as we can fall in love with a new person and build a relationship with our beloved, we humans have the capacity to fall in love with stamp collecting, bird watching, dogs, sports. … The list is endless and can be eccentric.

What we love — what catches our interest — is the basis of our learning because, as Doidge points out, love and plasticity work together. The toddler who points and coos when she sees a butterfly has an emotional connection to it — no matter how brief — and the attentive parent who notices that interest and responds to the child helps her learn something about this lovely creature she sees.

At first, the child just learns the name "butterfly" and perhaps learns the words to describe its colors. Who can say where that initial interest might lead? Some of us simply enjoy the sight of a butterfly. Others, later in life, will connect that pleasure with a poem, a painting or a piece of music. Still others will do research in lepidopterology.

The plasticity of our neural systems allows us to develop minds that are unique to the point of eccentricity. Some people's interests become so specialized and individualistic that they have trouble connecting with others. But the same plasticity that allows for such individuality allows for communication and cooperation.

It would be a mistake to equate the toddler's fleeting sense of awe in a butterfly with the kind of love that prompts the rewiring of the brain of a new parent. But, if nothing else, the connection between interest and learning is a revealing metaphor.

Many students who enter a university for the first time and start browsing the catalog of courses have an epiphany. They discover that the world includes people who love a wide range of subjects: medieval poetry, quantum mechanics, arachnids, electrical engineering — and those people have studied those subjects enough to earn doctoral degrees and to pursue research.

As the student begins, the opportunities for learning seem endless. The trick is to find the subject that the student really loves — a passion that will last a lifetime.

Chapter 5

The decline of the brain

Some of the best philosophical thinking about what consciousness is comes from physicians who have been forced to confront the loss of consciousness. While it's difficult to say what consciousness is, it is easier to talk about what consciousness is not. And, in the past 20 years, our thinking has changed with the discovery that consciousness is not an all-or-nothing proposition.

What happens to human beings who suffer traumatic brain injuries when it is not clear whether they are conscious? Are they still human beings? Are they persons?

Dr. Joseph J. Fins and Dr. Nicholas D. Schiff, who are colleagues at Cornell and co-direct the Consortium for the Advanced Study of Brain Injury, say that developments in neuroscience have shed light on those questions.

Change, technological and moral

In the past 20 years, we've learned to distinguish between people who are in minimally conscious states, rather than the vegetative states. While the term "vegetative state" seems to have been with us forever, the state was first described in 1972. Thirty years later, scientists described the minimally conscious state.

"People who are vegetative are not conscious," Fins said. "They don't feel pain. They are unaware of their environment. But their eyes are open. They have sleep-wake cycles. A minimally conscious patient may look most of the time like a vegetative patient — but they *are* conscious. They *are* aware of the environment and of others and *can* feel pain."

Tragically, up to 40 percent of the cases involving minimally conscious states are misdiagnosed.

Herbert case

One case strikingly illustrates the difference between the two states.

Donald Herbert, a firefighter in New York, suffered a disabling brain injury when a burning building collapsed on him in 1995. For almost 10 years, Herbert sat in a nursing home, virtually silent, almost blind. In an interview, Herbert's wife tearfully recalled consulting a specialist, searching for a shred of hope. The specialist told her that there was nothing there.

Then one day Herbert came back.

"How long have I been gone?" he asked. He wept when he was told.

Still unable to see, Herbert recognized the voices of family members and friends. When Herbert heard the voice of his youngest son, he was startled that he was no longer a baby. In his mind, time had frozen. When Herbert was injured, the boy was barely more than an infant. Herbert was shocked at how much his son, then 13, had grown.

It's a touching story. But try, for a moment, to look at it through the eyes of a scientist. How did those memories survive in such a badly injured brain? Even when Herbert was unresponsive, something was there.

Awakened by a sleeping pill

Schiff, a neurologist whose work has been recognized by *Time Magazine* and *Rolling Stone,* said that a diagnosis distinguishing a minimally conscious state from a vegetative state is critical because each has "a very different natural history pattern of recovery." Many minimally conscious patients can continue to recover for years. Some reach a level of independence and find themselves able to work.

In a famous case, a Texas man named George Melendez was in a minimally conscious state after almost drowning. He lost the ability to speak. But because he was groaning at night, his mother gave him Ambien, a common sleeping pill.

Melendez started talking and, with time, was able to feed himself. But he still needs the drug to be able to feed himself.

Schiff said the case is important for two reasons.

"It is a very clear example of how much brain function the severely injured brain may continue to harbor without evidence at the bedside of this latent capacity," he said. "It also shows us that after severe injuries it is often necessary to reorganize the dynamic activity of the brain to regain more optimal function, much like a cardiac pacemaker restoring the 'pump' function of the heart."

Deep brain stimulation

Schiff spent years developing a systematic model for such a brain pacemaker approach and then led a team to test this investigational therapy for the first time on a man who was severely beaten in a mugging. Most of the time, the man was unresponsive, but brain scans revealed that he did recognize his mother's voice. Schiff's approach employed deep brain stimulation — a technique used on people who suffer from Parkinson's disease that involves implanting electrodes in different places in the brain. The region Schiff's team targeted was the central thalamus. The treatment was successful, and the patient's mother spoke of her son's coming back — he regained his ability to eat, to stay awake during a conversation, to express pain and emotions, including love.

"This case tells us that we can intervene late in the course of what appears to be a stable and permanent level of recovery and make a significant functional difference," Schiff said. "We found that deep brain stimulation produced not only these effects, but it initiated an ongoing process of recovery despite a six-year period of lack of change without the intervention. The findings also confirm a strong prediction about the critical role of the brain network we stimulated in the recovery process."

Civil rights issue

In his latest book, *Rights Come to Mind: Brain Injury, Ethics and the Struggle for Consciousness,* Fins argues that the fact that four out of 10 minimally conscious patients may be misdiagnosed is a profound moral and legal issue.

"People in minimally conscious states, if they don't progress quickly, end up in nursing homes, where they typically don't get the standard of care they deserve," Fins said. "They are ignored and they are segregated — and I use that word in the context of the Civil Rights Movement. They are getting separate and unequal care. Their rights are being violated. The Americans with Disabilities Act says that people should not be segregated and should be maximally integrated into society. We understand what that means: we improve the sidewalk to help a person in a wheelchair get to work. But we don't often think of people who are conscious but minimally so and the assistive devices that could help integrate them back into society. That's their ramp back into the nexus of their homes and families. Think of the indignity of being in a nursing home bed, being ignored, being turned without pain medication because everybody thinks you're not there. … I think it's the next wave of the Civil Rights Movement."

Fins, who describes himself philosophically as an American pragmatist, quoted John Dewey's remark that technology often changes the moral and ethical landscape. Our understanding of consciousness, advanced mainly by changes in imaging technology, has changed the moral landscape, Fins says. While it's still difficult to define consciousness, it's not at all difficult to be outraged by the treatment of people who retain some consciousness, such as an awareness of pain, when they are presumed to have none at all.

"It bothers me that people who are conscious are treated so badly — as if they were not there," Fins said. "The brain is resilient, and people do recover. They deserve a chance. It seems to me wrong that we spend $600,000 to $800,000 to get people out of the hospital after a serious brain injury, and then we don't spend money to make sure that they can return to some semblance of a life. As a scholar and a scientist I'm dispassionate. But I think it's a moral wrong, and I want to right it."

Worthen case

Fins' book tells the story of Maggie Worthen, a senior at Smith College who suffered a catastrophic brain injury (Joseph J. Fins, *Rights Come to Mind,* Cambridge University Press, 2015). Maggie played Frisbee, spoke Spanish and wanted to be a veterinarian.

After her injury, Maggie was minimally conscious. Fins and his team helped Maggie to communicate. With rehabilitation — and the reorganization that was occurring in her brain — Maggie learned to use her left eye to answer yes-or-no questions. Later, she used glasses that could detect eye movement.

"But it wasn't so simple," Fins said. "These glasses are not routinely available for those with brain injury. One of our graduate students had to go to Radio Shack and rig up a pair. Sadly, technology that is widely used in the gaming culture wasn't available to people with serious injuries. It makes me wonder about our priorities."

Wallis case

Another famous case involved Terry Wallis, an Arkansan who was injured in a car accident in 1984. Wallis was thought to be in a persistent vegetative state.

In 2003, he started talking. Among his first words were "mom" and "Pepsi." Within weeks, he had greater fluency. One of the curious aspects of his case was that Wallis had been locked in time like Rip Van Winkle. To him, Reagan was still president.

For years, the medical profession had no category to describe Wallis's condition.

Wallis's mind was getting better, while his body, in general, was deteriorating.

Cases like Wallis's illustrate the possibility of undiagnosed covert consciousness.

This was illustrated in a paper by Adrian Owen, a scientist at Cambridge University, who used neuroimaging to identify people who appeared vegetative but had covert consciousness. Using magnetic imaging technology, Owen studied the areas of patients' brains associated with motor activity and spatial recognition. He asked patients to imagine themselves playing tennis. Those who were minimally conscious showed brain activity in regions associated with motor skills. Owen then asked patients to imagine themselves walking through their houses and studied the images associated with spatial reasoning.

Schiff and Fins and colleagues did a similar study of Maggie Worthen but since she had been a swimmer they asked her to imagine she was swimming. As Fins has observed, "Although Maggie didn't make waves, results like hers present a sea change in how we think about patients at the bedside. What they appear to be may not be what they actually are — covertly conscious." And there is a therapeutic angle too. Neuroscientist Martin Monti has used the scanner to ask patients, who could not otherwise communicate, to respond to yes-no questions by using a simple code. A patient might be asked to imagine himself playing tennis for a "yes" answer, for example. The answer would come in the form of pattern on an image of the patient's brain. While this is not yet ready for the clinic, it is, as Fins notes, "a promissory note for the future."

Scientific progress, clinical neglect

Fins describes the era as one of scientific progress and clinical neglect. "The health-care system doesn't think this is interesting," he said.

Fins said two things might be contributing to the inertia.

First, physicians were, for

generations, trained to think that brains never recover from injury. In fact, they sometimes do.

Second, training has emphasized the importance of making prompt decisions to withhold or withdraw care and to preserve organs for donation.

Fins sees too many "premature palliative recommendations — all despite stunning scientific progress."

Why such resistance to change? Perhaps the answer lies in the development of bioethics, which since the 1960s has asserted a negative right to die. It's essentially an argument for self-determination, the right to end biological life with the end of cognitive or sapient life. The cases of Karen Ann Quinlan, who died in 1985, and Terri Schiavo, who died in 2005, captured the attention of the nation.

But our understanding of the brain has changed. In the 1960s, the doctrine on traumatic brain injury was that nothing could be done. That view has changed with the understanding of the minimally conscious state, which offers, in some cases, a platform for further recovery. As Fins noted, "the presumption of futility appropriate to cases like Quinlan and Schiavo were generalized to other patients who appeared vegetative but were not, leading to therapeutic nihilism."

Fins argues that the change in our understanding changes the moral landscape. At a minimum, he said, we should do four things:

• Get the diagnosis right. That's impossible with a single visit from a doctor. Brain states can change. "We have to make sure they are credibly diagnosed so that families are making informed decisions about what their choices are."

• Improve access to rehabilitation. Too often, patients are warehoused. We ought to worry about isolation and pain in sentient beings, who are mistakenly thought to be vegetative.

• Provide families with respite care. He sees a role for hospice organizations to support families that "are in a state of almost constant anticipatory grief or bereavement."

• Educate families and the public. It's a good story, Fins said, "a story about resilience and the power of neuroscience to change lives." Unfortunately, it hasn't reached a broad enough audience.

Facing decline

What will happen as great numbers of America's 70 million Baby Boomers are diagnosed with Alzheimer's disease? That number has topped 5 million and is rising.

Cancer is a feared disease, but more people fear Alzheimer's. They fear the loss of control, the loss of the ability to make decisions about their own lives. How do you address that fear?

While some people argue for a right to die and are demanding changes in laws on euthanasia, that's still a minority view in the United States. Most followers of the major religions question whether a person has the moral authority to end his or her own life.

Moral decision

Paulie Gaido, author of *Mapping Your Own Alzheimer's Journey,* is sensitive to those religious values. Gaido, an attorney who has advanced degrees in theology and medical humanities, points out that Alzheimer's disease is a moral issue, as well as a legal one. The disease is a moral issue because it has a known trajectory, which means we can prepare for it. Failing to prepare for it is itself a moral decision.

The trajectory of Alzheimer's disease is well known. It's progressive, meaning that it slowly destroys memory and cognitive abilities. As a practical matter it's enormously difficult, in any one case, to find a precise point where a person is no longer able to make decisions.

And of course the disease progresses beyond that point. Sooner or later, a person suffering from Alzheimer's disease is likely to be unable to swallow food and water. If you are one of those who believe suicide is not a moral answer, what do you tell your physician? What would you want done when the disease progresses to that point?

Deciding for yourself

Since Alzheimer's disease and other forms of dementia erode a patient's ability to make decisions, it's important to leave instructions to physicians in advance — to make a written record of your wishes while your mind is still good. Making sure your physician has those instructions, known as advanced health care directives, is perhaps the greatest kindness you could show your family and loved ones.

"Your serious illness is traumatic enough for your family without forcing them to make medical treatment and care decisions on your behalf, should you be unable to do so," Gaido said. "Taking the time to express, in writing, your medical treatment and care preferences is, in my opinion, a good way to say 'I love you and don't want to burden you with the responsibility of making decisions on my behalf without adequate guidance from me.' Telling your spouse: 'You'll know what to do' or 'I am comfortable with whatever decisions you make' significantly underestimates the difficulty in making decisions, especially when the cost of medical treatment isn't completely covered by insurance, or has significant side effects, yet is capable of sustaining your life."

It's obvious, practical advice. Relatively few of us follow it. Few of us like to think about the possibility of sustaining an injury or suffering a disease that would incapacitate us, leaving us unable to make decisions for ourselves. To get past that reluctance, Gaido recommends starting with a statement of your values — the things that define you as a person.

"Advance directives typically communicate the *what* of your medical treatment and care decisions. In order to maximize the moral authority of those decisions you need to communicate the *why,* especially that those decisions are anchored in, and totally consistent with, your values."

Two vital steps

Gaido has boiled his approach down to two key points:

First, be honest. Tell your loved ones, as well as your physician, what you want done.

Second, appoint one person to make decisions on your behalf, just in case the actual circumstances of your disease or injury are significantly different than you anticipate.

"Your medical treatment and care decisions, even when made in advance, have undisputed legal authority. Without corresponding moral authority, however, those decisions could, under certain circumstances, be ignored or simply overridden by those who are convinced that they know better than you do what is best for you."

Chapter 6

The death of the brain

A little more than 50 years ago, scientists, physicians and policymakers pushed to change the definition of death. They argued that advances in science had allowed us to see clearly that when the brain was dead, life had ended, even if the heart was still beating.

A committee set up by Dr. Henry Beecher, an anesthesiologist and a professor at Harvard, began to work on a better definition of coma — a roundabout way of getting at death. For centuries, life had been equated with a beating heart or with breath. The Beecher committee argued that advances in science had shown that equation to be simplistic. A heart can be beating, but if the brain has died, a person is dead.

The committee's report prompted a cry that a new statute was needed. A commission set up by President Carter and continued by President Reagan produced a definition of death that was incorporated in the Uniform Determination of Death Act of 1981. It says that anyone who suffers "irreversible cessation of all functions of the entire brain, including the brain stem, is dead."

The law became standard across most of the United States, Canada and Europe.

The law changed what it meant to die across most of America.

Dan Wikler, a philosopher at Harvard, drafted the chapter that explained to people why the change was needed — why a person whose heart is beating but whose brain is dead should be considered dead.

Transplanting organs

The new legal framework was crucial to a new medical technology: organ transplants. At the time, the conventional wisdom was that the American public would accept the removal of vital organs, such as the heart and lungs, only from a person who was dead.

The practical problem was obvious. Organs are viable for a limited time after the heart stops beating. Allowing doctors to remove organs from donors who were brain dead, but whose hearts were still beating, greatly increased the number of available organs.

The effect of the Uniform Determination of Death Act and the other laws it inspired was profound: Countless people owe their lives to transplants.

Consensus challenged

In the decades since then, the definition of brain death has been controversial but at the same time widely accepted, at least in the Western world.

But in December 2013, the case of Jahi McMath, a 13-year-old girl from Oakland, Calif., challenged — and some would say "unraveled" — the consensus behind the definition of brain death.

Jahi, who had sleep apnea, had surgery on her airways. She

hemorrhaged, and her parents were told that their daughter had been declared dead. Her heart was still beating, but the declaration was made using a standard test for brain death.

Jahi's parents objected. The hospital replied that it didn't need their permission. Determination of death was not the parents' call.

An attorney who volunteered to help the parents found that in New Jersey, the law gave parents the right to decide whether the old definition of death or the new one applied. That provision in the law had been added to address the concerns of some Orthodox Jews whose religious beliefs equated life with breath, rather than with brain function.

And so, with the Super Bowl as a distraction, Jahi was taken, without the hospital's approval, by ambulance to a Catholic hospital in New Jersey.

Jahi's medical condition did not change. But, in terms of the law, a miracle had occurred: Jahi had come back to life.

Legal limbo

Jahi's mother and other family members went through an agonizing period. Their daughter was dead in the eyes of the law in California but alive in New Jersey. They received threats from the federal government saying they were claiming Medicaid benefits on behalf of a deceased dependent. They received mail from outraged taxpayers.

The conflict in statutes also confused the family's legal claims. A living person who was badly injured while in medical care can receive far greater compensation than the family of a dead person.

Years passed. Jahi went through puberty. Family members claimed that Jahi could understand them.

Jahi's mother posted many videos from her cell phone. She would ask Jahi to move her hand or foot. She'd ask Jahi to move a finger.

Wikler, who saw the videos, found them disturbing.

"The battle was joined — even among neurologists," Wikler said. Some thought she was minimally conscious. Others did not.

Brain scans found that much of Jahi's brain was intact.

Tests for death were supposed to be certain. What had happened here? Was this a single, remarkable exception to a sound scientific principle? Or had something gone fundamentally wrong?

National debate

The debate reached a national audience when Jahi's story was told by Rachel Aviv in an article titled "What Does it Mean to Die?" in the New Yorker on Jan. 29, 2018.

In her family's view, Jahi McMath died on June 22, 2018, four and a half years after she was declared dead.

Wikler tells the story as a clash of two views of death: the traditional one based on heartbeat and breathing and the newer one based on the functioning of the brain.

The unresolved debate raises questions at two levels.

First, are the tests used to determine brain death *always* accurate? Many people who supported the brain-focused definition thought they were.

Second, does being brain dead make you dead?

Jahi McMath's case seems to challenge the consensus answers to both questions — and of course the second question is fundamental: Was she dead really?

Second thoughts

Wikler, who was among those involved in the original push for a new definition 50 years ago, has had second thoughts.

The Beecher committee report had an ulterior motive. As long as there was legal cover, doctors could save huge numbers of lives with transplants. Physicians realized they were in a new era in terms of science and technology. Medicine had evolved to a point where it was keeping many people alive, but they were never going to wake up.

The potential for transplants was huge, but the committee did not mention that fact. Even the committee's name suggested that it was investigating a new definition of coma, rather than of death.

Coma dépassé

Before the Beecher committee convened, French doctors and scientists had described a medical condition they called coma dépassé. The state was similar to being in a coma, except that no one ever came out of it.

Crucially, in terms of the Beecher committee's work, coma dépassé came with a simple test. In one version, ice water is squirted into the ears and the physician looks for eye movement. It was billed as being a test that offered complete certainty.

Coma dépassé might have been a footnote in medical literature had it not been for a conundrum with the American view of vegetative states. The vegetative state is different from the persistent vegetative state, from which there is no return. But there is no test for the persistent vegetative state. It applies only retrospectively.

And there are cases where people do wake up.

Why did the Beecher committee focus on coma dépassé? It came with a relatively simple test. It allowed someone to say: If you're in that state, you're dead.

Wikler said the Beecher committee report was issued as if there were no ulterior motive — as if the potential for a boom in life-saving transplants were no consideration. It was presented solely as a matter for making use of new science.

He also said the report was issued without addressing the question of consciousness. And, as the McMath case shows, that question is critical to families. Members of Jahi's family thought she was conscious of some things. They also thought she might regain consciousness.

A continuum of cases

How *should* we define "death"?

Wikler contends that it's helpful to consider cases that fall along a continuum:

- A cold corpse.
- A person declared brain dead, but who is still breathing with a

respirator.

- A person, like Jahi, who has been declared brain dead but who might be conscious.
- People like us, who are reading books and discussing difficult ideas.

Wikler's point is that we are sure about us and the cold corpse, but we don't know about the intermediate cases.

In coma dépassé, the condition can persist for months. People continue to breathe, grow hair and fingernails. Young people can grow in height and mature sexually. Women in coma dépassé can gestate a child and bear a healthy baby. Is *that* a corpse?

A contradiction

Wikler outlined a thought experiment that suggests a

contradiction in the reasoning behind the existing law.

Suppose that a famous poet depends on a breathing machine, the result of a case of childhood polio. Despite his handicap, the poet is admired for his work, which has the capacity to move countless people.

Now suppose that the ceiling caves in on the poet, and the accident leaves him permanently unconscious. If the only reason we now say he is dead is that he is on a respirator, we have a contradiction.

Wikler contends that the argument that brain death *is* death is unsustainable. The question about when death occurs might have more than one answer.

The more productive question might be, "But when is *my* life over?"

Many of us would say this: "When the parts of the brain that support memory and continued consciousness are gone, I'm no longer present. My remains might be alive. But I am gone."

That question is more productive simply because it can be answered. While we might disagree on the answer, each can answer for himself through an advanced directive.

Part II
INTERDISCLIPLARY LEARNING

Chapter 7

Learning without borders

Our knowledge of the human brain has undergone a revolution. This new knowledge ought to inspire us to reconsider our own views about how we humans use our brains — how we learn, not only as children but over the course of a lifetime.

Many of us grew up under an educational system that assumed we all have individual gifts. The thinking went like this: Some kids are just smart and can learn better than others. Also, some kids are naturally smart in certain fields. Some kids might be naturally smart when it comes to math. Others might be naturally smart when it comes to languages or art.

But if we've learned anything about our brains, it's that they don't really work that way. Our brains are not fixed, so they don't have "gifted" areas that are fixed and immutable. Our brains are flexible and adaptable, capable of making new connections around obstacles.

The old conceptual model of how education should work doesn't align with the facts. The fact that the human brain is remarkably adaptable suggests that we are just beginning to tap its potential to learn. Put another way, we ought to revolutionize our thinking about education in the same way scientific discoveries have revolutionized our thinking about our brains.

How we learn

Rather than thinking about smart kids with innate gifts for art or chemistry, we ought to think about interests. Babies have them. Toddlers have them. Teens and senior citizens have them.

Perhaps only a poet, rather than a neuroscientist, can tell us why one toddler points to a butterfly, completely enchanted, while its twin becomes enthralled by a frog. But we now know that when a young child expresses an interest, the thing to do is encourage it. The brain is making connections. That process can and should be encouraged.

The experience of becoming interested in something occurs throughout life. We might not all have a profound interest in algebra at the moment that the math teacher convenes class on the first day of junior high school. But some people, regardless of age, run across puzzles that capture their interests, and they get interested in mathematics as a way to solve those puzzles.

Sometimes, the interest in mathematics becomes *the* interest. Sometimes, the interest in mathematics is just a tool in pursuing a deeper interest in another subject. But interests constantly evolve, and the brain is flexible enough to make new connections, accommodating the shifts in interests. It's not that the precocious first grader has a "gift" for ornithology. Rather, she's interested in the crows that live at the back of the orchard. She wants to learn everything she can about them.

From ants to Ivy League

E.O. Wilson, arguably the most famous biologist of the day, has written about the importance of following your interests. Wilson, who taught at Harvard, started with a boyhood interest in insects, particularly ants. He wondered why some insects are so highly developed socially while others are not. He followed that interest his entire life.

A basic interest in insects led him to other interests in other disciplines that broadened his understanding of insects. You can make a hobby of collecting insects. But to really understand them requires an understanding of biology, chemistry, physics and mathematics.

Wilson has argued that the proper course in education is to encourage that initial interest — in ants, in his case — and let the interest grow naturally into the other disciplines. The wrong thing to do is what we in fact usually do: tell the aspiring scientist that she cannot study insects until she has mastered the prerequisites.

We ought to encourage an individual's interest and let each student see where it leads. Of course there are exceptions. But typically we learn by following our interests, not by following the abstract demands of a curriculum.

Crossing disciplinary boundaries

The example of the young entomologist illustrates two important points:

First, learning begins with a simple interest. Just scroll through the catalogs of courses at a major university and you'll see the bewildering variety of human interest. It's hard to say why some people devote their lives to studying the capacity of bees to communicate while others study binary stars and still others study medieval Persian literature. But that extraordinary diversity of interest seems to be a feature of human beings. Individuals have their own interests. That intellectual diversity is a good thing.

Second, learning grows naturally across more than one discipline. This aspect of learning has been neglected, with the unfortunate result that we tend to confine education into specialized silos.

Humans are not really thinking until they get more than one perspective on a topic. Let's go further: The more perspectives we can get on a topic, the richer the thinking and the greater the possibility for understanding.

It is possible to learn in an academic silo. Some junior high students do learn algebra or geometry just by sitting in a class. But it seems that far more of us would learn those skills if we were allowed to pursue our own interests across disciplinary boundaries. For every child who starts out with a theoretical interest in mathematics there surely must be another who, like Wilson, learn math while pursuing an overriding interest in something else.

A plea to start early

Interdisciplinary learning is not just for graduate school. Many people can recall the first time — usually in grade school — that they had to make a report on an independent study. Professor Winslade, one the authors of this book, recalled that one of his teachers assigned him to write a report on the Panama Canal. And so, in grade school, he learned about the geography and climate of Panama and the history of the United States and Central America in the age of Theodore Roosevelt. One topic led to another. Learning about a canal obviously required him to learn about shipping, but it also required him to learn about railroads and the state of engineering at the dawn of the 20th century.

Often, we remember those experiences decades after leaving school. We were, at least briefly, excited about learning. We were doing something we loved. We got a chance to learn things we were interested in, but we also learned how to write a report and explain our interest to others — to our peers and to an adult or two.

The fact that those reports stand out in memory, at least for many of us, should give us pause. That fact implies that those brief periods of following our interests were the exception, rather than the norm.

Such a report is just one way to encourage interdisciplinary learning. The seminar, as opposed to a classroom lecture, is another way to encourage interdisciplinary learning, and it ought to be for elementary kids as well as grad students. The seminar allows for a back-and-forth exchange of what each student learned on his or her own. It allows students to ask questions and perhaps have more than one "teacher" giving a view. (Perhaps a student who was assigned to make the report acts as a second teacher.)

During the Moody Lecture Series, grade-school and junior-high students from Ambassadors Preparatory Academy formed a Brain Team. Students on the team attended all the lectures, and each kept a research notebook, studying the research interests of each speaker. Invariably, speakers in the series asked, "Who *are* those kids?" The speakers were impressed by the quality of the questions and the interest that the students had in neuroscience.

Public opportunities

The lecture series was itself an example of interdisciplinary learning. The series included all kinds of ideas from different perspectives. Among the speakers were people with expertise in neuroscience, psychology, philosophy, neuroimaging, medicine and education. The audience also was diverse, and the questions reflected the interests of students and faculty from a medical school, biology students from local colleges, retirees and public school students.

Those who attended got a better understanding of the brain, mind and consciousness.

Professor Winslade, who organized the series, said: "The lecture series was for me as much as the audience. When the series started, I did not know what people were going to say. But people who attended got into it, asked questions, got engaged. That's a good thing. The series awakened our interest in interdisciplinary learning. We know it goes on in different places in many ways. But what we realized is that if you want to learn something that's complex, you can't do it from one perspective, one point of view. Learning about the brain requires that you learn something about physics, chemistry, biology, philosophy, psychology and education."

And, when you think of it, don't almost all the really interesting questions have to be addressed that way? Don't they require more than one perspective, more than one point of view?

Chapter 8

Cultivating your passions

In the last chapter, we tried to make the case that one of the practical lessons we should learn from all these advances in neuroscience is that human beings learn by pursuing their interests. If you're convinced on that point, it's time to tackle the next question: How does an individual discover what he or she is interested in?

There are two answers — both obvious, but each contradictory. In short, a paradox.

First, we find our real interests

naturally, just by common sense. Consider the freshman that arrives at a major university and has the option of taking hundreds of courses. How does she find her passion? It's a bit like entering a buffet line. The wise move is to take a small sample of everything before loading up your plate. In other words, sample a broad variety of studies before you commit to a specialized field. As you go, ask yourself whether you have found something that's mildly interesting or something that's sparked a passion that will last a lifetime.

The second and prevailing view is to forget about exploring the possibilities. Often, children are guided, counseled or pressured into choosing a respectable or financially rewarding path before they've had a chance to sample anything.

Child prodigy

We've all heard of the child who knew she wanted to be a doctor or a concert pianist by the age of 5 and who single-mindedly pursued her dreams. Some people know what they want to do as children. They get on an educational path that leads to a career, stick to it and pursue their goal relentlessly. If they're happy with those choices as adults, there's nothing at all wrong with that.

But there is something perverse about the notion that that's the common pattern or, worse, that it *ought* to be the pattern that fits everyone.

Consider just one counter example, the case of Ludwig

Wittgenstein, arguably the most important philosopher of the 20th century. He went to university before World War I to study engineering and developed an interest in aeronautics, then an emerging field. That initial interest in flight led him to the problem of designing a propeller. The task of describing the sinuous shape of a propeller precisely — that is, mathematically — sparked an interest in the foundations of mathematics, which in turn led to an interest in logic. That interest, in turn, led to an interest in the nature and problems of language.

That, in a nutshell, is the intellectual biography of one of the world's great philosophers. It's a story of how a person pursued his interests as they evolved.

Some people seem to know early

on what intellectual passion will govern their lives. But surely the more common case is that a person's intellectual interests change as his mind matures. Surely that person would be happier if he were given the freedom to explore those interests as they evolved.

Social pressure

Too many people are browbeaten into deciding the course of their life before they get to college. One of the great off-script moments of the lecture series was when David Van Essen, who led the team that published a revolutionary map of the brain, told elementary and junior-high students about his evolution as a scientist. He started out wanting to be a physicist, discovered his mind was not really suited for it, and shifted to chemistry. The problems in chemistry were intriguing — but were not for him the makings of a lifelong passion. He found that passion, oddly enough, when he visited a bookstore and picked up a book about the brain. He knew that's what he wanted to do with his life.

Van Essen was able to make enormous contributions to our understanding of the brain because of that passionate interest. It drove his studies, his research.

But too often the high-school student or college freshman is pressured to make a decision based on something else. Financial rewards always come up. Students are told that pre-med majors can expect to make so much money, while sociology and art majors can expect to make so much less.

That's a dangerous game to play. Making decisions based on something other than your own interests is risky. It's a recipe for a divided self — for a person who works at a lucrative job he's bored with and whose brain is occupied elsewhere.

Often, students are pressured to choose a field before they have a chance to check it out. That's particularly true of students who are under financial constraints and who are told they must go straight through school without wasting a dime on an unneeded elective. But, for many people, our initial interests just don't hold up. We've all heard of the business student who did well in college but simply wasn't interested in a life in commerce. Instead, he found his real passion in teaching math to children.

Following your passion

Once you find your passion, how do you cultivate it?

The age-old advice is simple: Find a way. Incorporate it into your work if you can. If you can't, do it on your own.

But to give up on your real interests, your real passion, is to go on with life without your brain — or at least without an engaged brain. Plato, in *The Republic,* talked about what it means for a person to be happy. In the dialogue, Socrates suggested that happiness reflects harmony in the soul. You are a happy person if you are happy with what you *do*.

Epictetus went a step further to say that unhappiness suggested coercion — a kind of slavery. A happy person is happy wherever he finds himself. When people are unhappy, they are doing something they don't like. The only reason a thinking person would do something he doesn't like is because he is not free to pursue a better path.

Tale of the three poets

How would you feel if your college-bound child announced that he planned to pursue a degree in poetry?

Consider three famous American poets.

• Wallace Stevens had just such a discussion with his father more than 100 years ago. Stevens's father, a lawyer, was appalled. He wanted his son to pursue a professional career, preferably law. Stevens rebelled initially but caved in and went to law school. He pursued a career in business law, eventually rising to vice president of the Hartford Accident and Indemnity Company. While he pursued a practical business career, his first love was poetry. His house in Hartford was two miles from his office, and Stevens walked to and from work each day, composing poetry as he went.

• William Carlos Williams is viewed as one of the great American poets, but he often said he had two loves — poetry and medicine. Art informed his medicine, and medicine informed his art. Williams got his medical degree from the University of Pennsylvania and practiced medicine in his hometown in New Jersey his entire life. As demanding as the medical profession is, it doesn't take up 24 hours in the day. Williams famously kept a typewriter on his desk and wrote between appointments. He kept a notebook in his car. He was known to have written poems on house calls, waiting for a baby to arrive.

• Mary Oliver took just the opposite view. She made a conscious decision not to pursue a professional career because it would interfere with her poetry. In fact, she made a decision never to take an interesting job. Instead, she worked a variety of jobs, often as a clerk in a shop. She said her employers always got her second-best effort of the day. Oliver knew that poets didn't make much money, so early on she made a list of all the things she would have to do without if she followed her dream. She made the list of things she'd have to leave behind and let them go. She organized her life around her writing and let practical matters take a secondary seat. In Oliver's view, abandoning your interests would make you unhappy, while abandoning a career would just make you poor.

These three poets are remarkably different — in voice, style and approach to poetry as a focus of their lives. But each was accomplished. The three have at least one thing in common: All won the Pulitzer Prize.

If there is a moral to this story, it's this: Don't be afraid to get off the script. Almost always, there's more than one script.

Changing your mind, changing your brain

If all this sounds a bit suspicious,

you're right to be wary. As Moheb Costandi observed in his book *Neuroplasticity,* the phrase "rewiring your brain" has become a pop culture phenomenon (Moheb Costandi, *Neuroplasticity,* Cambridge, Mass.: MIT Press, 2016). Self-help authors and New Age gurus have used the phrase in urging changes that they say will improve your sleep and work habits and rekindle your love life.

You're right to be skeptical that reading a single article or a book will result in changes in your brain that can be detected on your next MRI. But at the same time, it's now accepted that long-term training, especially if it begins when young, has an effect on the structure of brain.

Some of the early neuroimaging studies were aimed at the effects of bilingualism, which is common in Europe but rarer in the United States. Predictable changes could be seen in the density of gray matter in a structure, the left inferior parietal lobule, associated with language acquisition and use (Costandi, pp. 88-90).

Researchers also turned the imaging technology on to college students and military interpreters undergoing intensive training. Could you see a measurable difference in their brains? Sure.

Analogous changes occur in the brains of musicians and athletes. It takes a tremendous amount of control to be a concert pianist, for example. If you're looking at images of the brains of two 22-year-old piano students, you probably could infer which began to study at age 7 and which began a year ago. Similar kinds of studies have been done of novices and masters of karate.

One of the most imaginative studies was done on people who were studying to become London taxi drivers. To get a license, a candidate must show "the knowledge," a familiarity with 26,000 streets, some that follow the winding paths of ancient oxcarts, within six miles of Charing Cross Station. Suffice it to say that such an undertaking does things to your brain — particularly to the structures that are involved in spatial navigation (ibid., 94).

The point here is that changes to the brain occur over a long time. A young man who decides to become a weightlifter does not see much visible change after his first session in the gym. After 10 years of regular practice, you'd expect to see change. Similarly, people who engage in something they're passionate about — who work at a subject for hours a day for 10 years — are doing something that is changing the structure of their brains.

We all have fleeting interests — an interesting article in a newspaper, a bit of conversation at the coffee shop. Perhaps each changes you in some small way. But sustaining interests, the kind of interests that become passions that last a lifetime, change you in ways that are discernible.

That evidence ought to bolster the case that the chief mission of any educational system is to help a student find and cultivate her own interests.

Chapter 9

Independent learners

I hold that so long as I live, I must have a student's inquiring mind and a thirst for learning. — Gandhi

Interdisciplinary learning is a good way to develop independence. We're talking, not about financial or political independence, but something more fundamental: intellectual independence — the ability to think for yourself.

It's possible to develop that capability by following the traditional route, by choosing an academic discipline. But, if you're interested in developing your own mind, in finding out what you think, as opposed to simply absorbing the thoughts of others, then cutting across disciplines offers an obvious advantage. By definition, interdisciplinary learning involves more than one perspective on a topic or problem. You can't have an argument — not even in your own mind — without having more than one point of view to consider.

Many people of a certain age will roll their eyes at the "compare and contrast" question, once the standard of the high school essay. But comparing and contrasting different points of view is a step toward forming your own view.

Academic work generally

involves a review of the literature. You find what others have said and begin from there. Interdisciplinary learning takes that process a step further and considers the literature from more than one field.

A few examples

This type of learning might be difficult to define, but you can get a suggestion of how it might work by looking at some examples of interdisciplinary learners.

Interdisciplinary learning is something you *do* — it's an activity, rather than a concept or curriculum. It's an *approach* to learning that can take many shapes.

Rather than tackling interdisciplinary learning as a theory, you can get a better grasp of it by looking at people who embodied it — who were living examples of interdisciplinary learning. Here are some of our favorites.

Richard Wright

Richard Wright is one of America's great writers, and his autobiography, *Black Boy,* is one of the great declarations of intellectual independence. Wright argues that you're wasting your life if you're trying to conform to the standards of other people.

Wright grew up under two powerful influences: the racism of the Jim Crow South and the religious beliefs of his family. Those in political power told him he must behave as if he were less than human. Family members told him he was not free to choose his own beliefs and values — that his family could and should dictate them.

We humans spend our teenage years trying to emerge from the cocoon of family to become independent adults. Wright's story can have an astonishing affect on those teenagers who suspect that their spirits are being crushed by unhealthy forces.

Wright's message is that it's your responsibility for finding your way in life. If you let social forces put you in an unhappy place, they will. It's up to you to inform yourself, to educate yourself. If you can get help, fine. But if you can't, you've got to do the best you can on your own. You can't abdicate that responsibility. To give up on forming your own mind is to give up on life. You have got to pursue your own interests, learn about the world, other people and yourself.

In Wright's case, the social forces were violent and cruel. He was prohibited from getting an education by a culture that wanted African-American boys to grow up to be nothing other than a source of cheap labor. The efforts to keep African Americans from thinking independently are almost unfathomable today. But Wright didn't let himself fail. He describes how, when he was prevented from borrowing books from the library, he did so through ruse and trickery.

Wright escaped the Jim Crow South and continued to educate himself. He spent his life as a writer exposing oppressive social and political forces.

In a way, he is the most American of writers. His voice urges people to live by their own lights.

Abraham Lincoln

Have you considered the educational qualifications of the man who is widely regarded as America's greatest president? They are modest. No college degree, little formal schooling.

Further, when experts in childhood development refer to "hostile parenting," they might have used Lincoln's father as a model. Thomas Lincoln was illiterate and considered education to be a waste of time. He actively discouraged his son, but Abraham became a learner anyway. It's hard to grasp the fact that the author of the Gettysburg Address did not grow up speaking standard English.

Abraham Lincoln realized at an early age that public speakers needed to know something of grammar. So he borrowed a book and worked his way through it.

He had to work slowly to get the principles. As he studied speech, Lincoln began to think about how people are persuaded. He saw, from studying examples, that a good story or illustration is often better than an argument. And then he began to think more about what people should be persuaded about — things such as the value of public institutions and common justice.

Lincoln studied law the same way he studied rhetoric — one

borrowed book at a time. It was the pattern of a lifetime. He read and learned about things that interested him. On the night he was assassinated, his pockets were full of newspaper clippings.

A book of your own

Sooner or later, you'll run across a mentor who suggests that everyone should write a book. This mentor is talking about a personal book, a book that reflects your own thoughts and opinions, your own view of the world.

Michel Eyquem de Montaigne wrote such a book almost 450 years ago. He started small, trying to record his own ideas about some of the questions that human beings keep coming back to — what it means to get an education, to love, to live honorably and to die a good death. As he thought, he wrote. The book grew, question by question, topic by topic. The work always reflected Montaigne's own viewpoint. And so what emerges is the story of a person, and a personality. His book is a record of an individual trying to make sense of the world.

Montaigne's book — he called it *Essays,* meaning "trials" or "attempts" — is one model for this kind of enterprise. Walt Whitman's book is another.

Whitman began *Leaves of Grass* as a collection of 12 poems. It grew as Whitman grew. Whitman collected experiences as a child might collect stamps. He often got on the omnibus, sitting next to the driver, just to hear what people on their way to work or to market were talking about. He went to hospital wards to see the cost of the Civil War. Each day, Walt Whitman went out into the world, looked around and reported on what he saw. And so the book grew.

One of the many benefits of writing your own book is that it reminds you to be aware — to pay attention to the people and things around you. If you are going to compile such a book, you have set yourself a task: you must go out each day to see and appreciate the world's wonders and to bring one small piece back.

This kind of book admits many forms. Montaigne wrote essays. Whitman wrote poems.

Antonio Porchia wrote aphorisms. Porchia was a poor man, living in a small house in a suburb of Buenos Aires. Instead of writing essays and poems, he tried to distill the experiences of the day into aphorisms. He called his collection *Voices.*

Though these three learners wrote in different forms, each devoted his life to one work, one collection of his own thoughts.

Sheilah Graham: College of One

When Sheilah Graham, an Englishwoman, came to America and fell in love with F. Scott Fitzgerald in 1937, she wanted to get an education. Fitzgerald designed a two-year reading program for her. She called it her "College of One," and published a book about her experience using the phrase as a title.

This quest started as she was trying to read Proust. Fitzgerald recommended 10 pages a day for the first third of the book, 30 for the second, and 40 for the third. It's the advice a professor at the University of Texas used to give students learning to read Ancient Greek: The more you read, the more you *can* read.

The next step in Fitzgerald's plan was a list of classic plays and novels, tied to 25-to 30-page sections of H.G. Wells' *The Outline of History*. The readings didn't relate. The point was simply to get the student to absorb history and literature at the same time.

Fitzgerald believed in short courses — three weeks for a history of France and six weeks for music, including recordings. But those "courses" would come later.

Part I of the College of One curriculum was 40 history lessons, with 40 works for literature sandwiched in between. This approach has obvious limits: Fitzgerald's program included no math, no science, no languages. But it's a way for a person who has had a limited education to get a better grasp of literature, history and philosophy.

Graham's College of One is perhaps the simplest of many approaches to lifetime learning, an idea that was perhaps most popular in the 1930s. The University of Chicago adopted the Great Books Program, in which students read the original text of great thinkers in the West, starting with ancient texts, including the Greeks and Romans, and adding scientific thinkers, such as Galileo and Newton, as mankind developed a scientific view. The curriculum is still used at St. John's College.

Reading plans

Clifton Fadiman, a scholar associated with Columbia, brought the idea to the masses with his book *The Lifetime Reading Plan,* which suggested a course of guided reading through the Western canon. Many other scholars, including Harold Bloom of Yale, have offered similar suggestions for getting an education by reading.

A more recent spinoff to the Great Books idea comes from Paul Marcantz Cohen, dean of the honors program at Drexel University, who makes a case for one-credit classes. Instead of rushing through *David Copperfield* as part of a larger course, students read a single great book. The students get one semester hour of credit, rather than the usual three, for such a course. You can't really make a degree plan based on one-hour courses, but they do add some spice.

Cohen's point is that great books inspire people, including teachers, and inspiration is a good thing to have in any education. The program at Drexel started with books but now includes art, music, film and architecture. It gives students a feel for the experience of reading for pleasure, rather than just rushing through.

A similar idea is a 30-day course. Don't know anything about the poetry of Rita Dove or Sandra Cisneros? Find a book in the library and make a 30-day study of the topic.

Don't know anything about jazz? Find a list of recommendations of 25 great pieces of music, and spend a month listening and making notes. Not long ago, *The Guardian* newspaper, bemoaning the lack of interest in contemporary composers of orchestral music, ran a column on "25 works by contemporary composers."

Start in a field in which you are ignorant. You'll find you can learn something — perhaps not a great deal, but something — in a month. Perhaps that will be enough. But you might find that the taste piques your interest and inspires you to take a deeper look at the topic.

Finishing school

At age 70, Socrates took up the lute. His excuse: He was capable of learning a skill he hadn't mastered before. A human being is capable of learning. The idea of perfecting the person as he or she ages is as natural as the idea of fruit ripening.

Behind that idea is the notion that we humans are not finished: there is more to life than any one person has experienced. We feel the desire to be more complete at the close of the day than we were when we woke up.

At 81, Joseph Epstein, a master of the essay and longtime editor of *The American Scholar,* took up the study of Latin. Epstein said his early education focused heavily on memorization. When he moved to the University of Chicago, higher education was about finding intellectual connections between concepts. Learning Latin, he found, engaged his whole mind. It challenged his memory. It challenged his ability to connect ideas imbedded in complex sentences.

"The idea of a good time for many people my age is to do crossword puzzles, play bridge or Scrabble, hit golf balls," he wrote. "I prefer to wrestle with this long dead but still magnificent language." (Joseph Epstein, "Learning Latin," *First Things,* April 2020)

Eric Hoffer

More than one critic of corporate culture has described how the status of workers often descends into near slavery. We use our talents to achieve someone else's goals, earn profits for someone else's company. In our culture, the work increases until there is no room for the use of your own talents for your own goals. This is why Eric Hoffer is an American sage. He lived a life that showed the way out.

He started out by cleaning plates and washing dishes. He was a laborer, migrant farm worker and, when World War II came, a longshoreman. He took ordinary jobs — not those that were difficult to find and hard to get. He took what was available. The jobs did not pay well, but he did not have to worry about the supply of work. He started at the bottom and didn't move far.

He adjusted expenses to income.

He lived simply, renting a room, rather than a house. As a migrant worker, he carried his possessions in a knapsack. He worked through his notebooks, consolidating them, rather than keeping one after another.

What Hoffer really liked to do was read, talk, reflect and think. He worked in little notebooks or on note cards. He tried to get to a simple idea — a principle — and express it in no more than 50 words. He was as economical with words as he was with money.

Hoffer examined the ideas of his time. He was fascinated by the mass movements that led to Soviet Russia, Nazi Germany and Fascist Italy. He was interested in the New Deal that transformed the United States. How do human beings, who seem capable of living as individuals, get involved in mass movements? His investigations led to his most famous book, *The True Believer.*

Hoffer ordered his life in a way that he could conduct these investigations on his own terms, without interference. He avoided entanglements of all kinds — permanent jobs, marriage, housing.

His method is worth noting. He would go to the library and browse. He would see what interested him, rather than consult the literature. Collegiality — a feeling of community among scholars — implies a social order within that community. That notion is missing in Hoffer. Collegial relationships were not a part of his world or his life. He was driven by his own interests and he kept himself free to follow them.

Herodotus

You could make a case that Herodotus was the world's first great example of an interdisciplinary learner. The Polish journalist Ryszard Kapuscinski made that case in *Travels with Herodotus* (translated by Klara Glowazewski, New York: Vintage International, 2008). Herodotus was curious about differences — between Greeks and Egyptians, Persians and Ethiopians, Athenians and Spartans.

To find something out, you must inquire. And so Herodotus got on a boat — in the age when sea voyages were dangerous and pirates were many — and went to inquire. Journalists who bother to read him love him. Herodotus was the first really good example of someone who went about the business of reporting. Because he interviewed many sources, Herodotus ran across contradictory accounts. He did his best to weigh the evidence.

Herodotus was driven by his curiosity about other cultures, and he organized his life to satisfy that curiosity. He reported on other cultures' religions, farming practices, mathematics and wildlife. Before Herodotus wrote *The Histories,* the word "history" simply meant "inquiry" or "investigation." Herodotus' life and work are still wonderful examples of interdisciplinary learning conducted by an independent learner.

The well educated person

Since the Ancient Greeks at least, human beings have debated what constitutes a good education. Let's take one point of view that, flawed as it may be, has endured for more than 2,000 years.

Aristotle held that a well educated person studies a subject not to acquire scientific knowledge, but to become a discerning judge of the arguments. Specifically, a well educated person would be able to see, no matter what topic came up, which arguments applied to the problem and which did not.

The well educated person is able to inquire into all subject matters, not just one or two. He or she has the ability to judge across many subjects — ideally, all subjects. Aristotle was describing interdisciplinary learning.

Today, we live in an age of specialists. We need people who can bridge the gaps between specialists and their highly individualized specialties.

Over and again, we come to the same conclusion: that the really interesting questions facing human beings inevitably involve more than one specialty, more than one discipline. The lecture series on the brain, and the book that followed it, is a striking illustration of that point. To really get into the matter, you need to learn something about biology, neuroscience, imaging technology, psychology, medicine and philosophy. And that's just a start.

Chapter 10

Interdisciplinary learning, Part I

"The function of reason is to promote the art of life." — Alfred North Whitehead

Brains are *useful* things. Minds emerge from them, and from their interactions with bodies and environments.

The most *useful* way to train a mind is through interdisciplinary learning. Interdisciplinary learning is essential in solving problems. It encourages you to see different aspects of the problem. The more aspects of a problem you can see, the more options you have for solving it. You can't consider which of 10 options is best if you can see only one of them.

Training minds through an interdisciplinary approach helps students learn to be creative in solving problems. That's a far more useful thing to do with a mind than train it to memorize facts for the next standardized test.

The ability to make better decisions about difficult topics is useful regardless of the topic. An interdisciplinary approach can help you make better decisions, regardless of whether the problem you're trying to solve is a business problem or a problem involving your personal life.

Three illustrations

Perhaps that point about interdisciplinary learning can only be illustrated or shown, rather than argued. To illustrate how interdisciplinary learning allows people to solve complex, difficult problems, let's consider three cases that raise difficult ethical questions about the practice of medicine. All three cases were documented by Professor William J. Winslade, a consultant in medical ethics.

By using perspectives from a variety of disciplines, people involved in two of these cases were able to address problems with solutions that weren't immediately obvious. In one case, we'd argue, a tragedy was made worse by a failure to look at an agonizing problem from more than one perspective.

The case of Mr. T

The first case involved Mr. T, a colorful character who had been treated for problems with alcohol and who had been sent to a psychiatric hospital.

Mr. T, who retired in California, had been a taxi driver on the Isle of Man off the coast of Great Britain. He had a forceful personality and ruled the roost — his family was in awe of him.

Mr. T injured his foot, and gangrene set in. But he refused to have it treated. He came to the attention of the authorities when he fell asleep on the couch with a lit cigarette in hand. His conversation with firefighters prompted them to take him to a psychiatric facility.

The ethical issue was what to do about the gangrene. Two doctors said the foot had to be amputated. Mr. T refused. The concern was whether Mr. T — who seemed competent in some ways and incompetent in others — was competent to make *that* decision.

The case became a public matter when a family member contacted a probate judge. The judge realized the need for different perspectives. He appointed a lawyer for Mr. T and called on Winslade, who was chair of the bioethics committee for the local bar association, to act as an adviser to the court. Winslade hired a psychiatrist who was also a lawyer to find out why Mr. T didn't want the operation.

Mr. T expressed the belief that God would take care of him. He also said that a good walk in the salty sea would cure him. When asked if he understood that he might die if the foot weren't amputated, Mr. T was evasive.

It was clear that, while Mr. T was not totally demented, he was not firing on all cylinders.

During the court hearing, the judge listened to all perspectives. He then took a break. When he returned, he told Mr. T that he was instructing the doctors to amputate his foot.

Mr. T thanked the judge.

Everyone in the courtroom was flabbergasted. Mr. T had been uncooperative with doctors but had somehow come to respect the authority of the judge. The judge had two family members who had undergone amputations from war injuries and was familiar with the resistance to such surgeries. But he also understood the high likelihood that Mr. T would die if everyone failed to act.

Winslade thought Mr. T might have been consciously opposed to amputation but was unconsciously relieved when the judge ordered the life-saving operation. Winslade also thought Mr. T seemed to have been satisfied that the hearing had allowed the matter to be examined thoroughly — everything important had been said.

The surgery was successful, but Mr. T refused to cooperate with therapists who provided rehabilitation. Mr. T was sent back to the psychiatric hospital.

The judge, hearing that nothing positive had happened, reconvened the attorneys. The solution was pragmatic: Winslade found a nursing home where Mr. T's wife could come and go at will. The solution allowed Mr. T to get care and also allowed him to keep his family ties.

Winslade said a satisfactory solution to the difficult problem was found because the judge considered *all* the perspectives and then followed through. Finding the solution required more than a court hearing — it required an evaluation of the outcome and follow-up.

Learning, by itself, wasn't enough in Mr. T's case. Learning the facts and giving the right answer once, in a courtroom, wasn't enough.

Everyone knew Mr. T needed an amputation. Mr. T was not capable of making that decision for himself.

That much was learned at the court hearing. But the problem was how to enable Mr. T to resume a full life. The solution required further inquiry and second efforts.

About a year later, Winslade visited Mr. T to find out how he was doing. Although Mr. T was still uncooperative with therapists and was confined to a wheelchair, he was enjoying life, holding forth as a storyteller. By then he'd used his storytelling skills to give a new slant to his own case. When Winslade asked about his memories of the court hearing, Mr. T recalled that he'd gone to court to get *permission* to have his foot amputated.

Sidney Miller Case

The Sidney Miller case involved a tiny girl, born at 23 weeks with severe complications. It was widely covered in the news media as a "wrongful life" case.

The girl was born in a Houston hospital to Mark and Karla Miller. From the beginning, medical opinion was divided: the Millers' obstetrician and neonatologist disagreed about the best way to proceed. It was the 1990s, and neonatology was a developing field. The neonatologist wanted to deliver the baby because of complications. But the baby weighed a pound and was put on life support. Soon after, a respirator caused extensive brain damage.

The Millers were informed and tearfully decided that they did not want the baby kept on life support. Mark Miller left the hospital to make funeral arrangements.

But the neonatologist insisted the baby could not be taken off life support. The Millers did what they could, but it was clear that Sidney was not ever going to be able to function as an independent person.

The Millers sued the hospital, contending that its medical staff had taken actions in treating Sidney without the Millers' consent.

A jury agreed and awarded the Millers $65 million, including punitive damages.

The hospital appealed, and the Texas Supreme Court eventually ruled in favor of the hospital. Although the Millers won the jury trial, they got nothing.

Winslade described the case as a missed opportunity to examine all the options. The hospital didn't have to take the course of action it pursued. More than one perspective was available — there were conflicting medical opinions even at the trial — but the hospital stood behind one point of view.

In Winslade's opinion, the hospital's decision to follow the neonatologist's contention that the baby had to be delivered early and then kept on life support made a tragic situation worse.

Baby C Case

After the Sidney Miller case was tried, Winslade represented another couple that had a child in the same hospital. The couple had two children, both born prematurely. The first child had received excellent care and was thriving. But the second was in neonatal intensive care. As time passed the parents grew concerned, not sure what was happening.

A resident physician said the baby was not going to do well. It had a lot of problems because of premature birth. The parents were concerned, but they didn't want to cause problems because they relied on the hospital and its physicians for care for their other child.

The father of the child had two sisters, both of whom were physicians who lived on the East Coast. Both expressed alarm about the lack of information the parents were getting.

Winslade made inquiries on behalf of the parents and was told to back off. He then talked to one of the hospital's lawyers about having the baby transferred from Houston to the University of Texas

Medical Branch in Galveston for a second opinion.

If the baby was not viable, the parents did not want to keep him on life support. Winslade sent the baby's records to the chief of neonatology in Galveston, who concluded the baby had little chance of survival. However, the physician agreed to examine the child, and Winslade arranged a transfer.

The baby barely survived the trip. Physicians at the medical school agreed the baby had no prospects for a full life. The parents asked that the baby be removed from life support.

The parents were there, as was their older child and the father's two sisters. Everyone was sad, but all realized that there was no possibility that the baby would be able to function on its own.

Winslade contends the key difference between the Sidney Miller case and Baby C case was the awareness of alternatives. "By being aware of alternatives, we got a solution that was sad, but better than what happened in the Sidney Miller case," he said.

Seeking alternatives

These cases illustrate the process of interdisciplinary learning, which starts by trying to figure out all of the possible alternatives to a problem. The search for options crosses disciplinary boundaries.

Winslade took a position as a professor of medical ethics after having had experience as a medical ethics consultant. His approach was to try to find the possibilities that hadn't been considered.

The same kind of thing happens in interdisciplinary learning, whether you are looking at a problem in business or in law or in your personal life.

If you are in a situation where you are breaking new ground and you're trying to figure out how things should be handled, you need to first identify all the options.

When people make decisions impulsively, without considering all the options, they're likely to make decisions they later regret. Training in interdisciplinary learning equips people to make better decisions, not just in medical ethics, but in any field that calls for difficult decisions about complex problems.

Others' perspectives

The bias in our culture against looking at different perspectives is so strong that some seemingly difficult cases in medical ethics haven't been difficult at all — the solution was simply to find out what people who had been on opposing sides actually thought.

Winslade was once asked to consult on a case in which a woman entered a hospital to have surgery done to repair scars. During the operation, the anesthesia equipment malfunctioned, and the woman suffered severe brain damage. She was on life support with no hope of recovery.

She was kept alive as the legal issues were resolved. But when the litigation was settled, Winslade was asked to find out how life support could be discontinued.

He had to consult with various attorneys, courts and the husband. Everyone was cautious about discontinuing life support in light of the litigation. But it turned out that no one disagreed that it was the right thing to do.

Simply talking to people — finding out what their perspectives are — will sometimes solve a problem.

In a far less serious case, Winslade once was asked to inquire about a man who had surgery. The procedure was successful, but he was failing to thrive. He wasn't eating and regaining his strength.

Winslade simply went to talk to the man to see what the problem was.

The man replied that hospital food was terrible.

Winslade asked the man what he wanted to eat.

Brisket, he replied.

The problem was solved when the man's wife was allowed to bring lunch.

Chapter 11

Interdisciplinary learning, Part II

It would come as no surprise today to hear that a philosopher with an international reputation had published an essay warning of the "deadly" practice of teaching to the test.

But the essay we have in mind was published almost a century ago. Alfred North Whitehead, philosopher and mathematician, was talking about the folly of an educational strategy that continues to do harm today (Alfred North Whitehead, "The Aims of

Education" in *The Aims of Education and Other Essays*, New York: Macmillan, 1929).

In the early 20th century, educational reforms had given Britain the "uniform external examination," a standardized test. Schools, "on pain of extinction," had to prepare students for it, Whitehead said.

Whitehead found it ridiculous to think that an education was something that could be imparted by a battery of tests. Whitehead, a giant of modern logic, dismantled the idea.

Instead of teaching students specific information to pass a test, we should be trying to awaken the student to the *activity* of thought, Whitehead wrote. We should be trying to teach them to *actively* think for themselves, rather than passively accepting the great mass of ideas about the world that was handed down by preceding generations.

Ideas have to be tested, proved, *used,* he said. If not, they are dangerous.

This activity of thinking can be joyous, a point which we should celebrate and proclaim in teaching.

Whitehead asked a series of questions of would-be testers: How do you evoke curiosity? How, once that curiosity is lit, do you develop a sense of judgment? How do you encourage a student to have the patience to sort out conflicting evidence and look for new theories to explain the apparent contradictions? All those powers, Whitehead wrote, "are not to be imparted by a set rule embodied in one schedule of examination subjects."

The American system

For those who would enjoy watching a world-class mind ridicule a ridiculous idea, "The Aims of Education" is not to be missed. But if you're looking for a short answer to the question "What's wrong with the American educational system?" it's simply that it doesn't *work*.

The system encourages memorization. It doesn't encourage the kind of learning that results in students doing some work beyond that.

Too often, "learning" occurs when students store information in memory just long enough to take a multiple-choice test. The tests are graded, often by machine, and the results are recorded in the name of accountability. Before long, the information the students memorized is forgotten, but hardly anyone notices. Everyone has moved on to the next test.

The remarkable feature of this kind of learning is that it can't really be *used*. It doesn't prepare students to *do* anything with the material they covered and allegedly absorbed.

It doesn't train students to explore different perspectives on problems, analyze possible solutions and make decisions about the best approach. In other words, the prevailing system doesn't train students to draw their own conclusions.

A better approach would prepare students not just to memorize information but to *use* it to do work, such as writing reports, organizing projects, building models and solving problems.

An elementary student who does nothing more than memorize formulas is going to have a limited view of mathematics. But consider how much young students learn by making a scale map of the school grounds, calculating distances between each tree, fence, door and other feature. What kinds of mathematics are needed to ensure that the spatial relationships depicted by the map are correct? Students learn by solving problems, and that learning goes far beyond memorization.

There's nothing wrong with requiring elementary students to memorize some material. But if that's all those students can do, they are not getting much of an education.

Yet this notion of education, based on memorizing material to pass the next test, persists and has spread from the public schools to graduate schools. Today, law and medical students routinely take multiple-choice tests. Our system is obsessed with the ability of students, even at advanced levels, to memorize information. The ability to *use* that information seems almost an afterthought.

A better way

This problem has been discussed for years. The educational reforms of an earlier generation — no matter how well intentioned — have been disastrous, yet educators don't seem to be able to break the cycle of teaching to the test.

Perhaps part of the problem is that educators are looking for a new *theory* to counter the old one, which ran something like this: Our schools are in a mess because of a lack of accountability. To fix that problem, we'll test often to find out who is performing and who is not. We can then focus on the problems, i.e. problem students and problem teachers.

That theory has a record of failure in practice. We'd argue that a new theory won't do much to improve things. Instead, we'd argue for an educational *practice* — rather than a theory — that focuses on interdisciplinary learning.

In this view, education is not something that requires students to memorize things, that is, learn something *in theory*. Instead, this view of education stresses practices that require students to *use* knowledge to solve problems and to *do* things. In a word, this approach is pragmatic.

Common features

Since it's a practice, rather than a theory, interdisciplinary learning doesn't have a strict, formal definition. Instead, the many forms of interdisciplinary learning have some common features.

First, interdisciplinary learning means studying a problem from more than one perspective. That's the essential feature. Students need to see a subject or a problem from alternative points of view. It's a bit like looking at a sculpture. If you stand still, you'll get one point of view. If you move around the sculpture, you'll see it from many angles. You'll get a deeper appreciation of the art.

Second, getting different perspectives allows a student to see alternatives, and the ability to solve problems creatively depends on the ability to see alternatives. This is not a theoretical enterprise. It's a common activity that leads to more focused action. For example, it's a common practice to bring in consultants when an institution is facing a difficult problem. The consultants look at the problem from various perspectives because those perspectives provide them with different options for solving the problem. The different points of view are not just of theoretical interest, something to be memorized and recited at test time. Instead, each perspective suggests a possible solution. The consultant identifies those options and analyzes them, trying to determine which option — or which combination of options — is the best.

Third, this kind of interdisciplinary approach yields a more comprehensive style of learning. It's also a deeper style of learning. If you approach a complex problem, the most obvious solution is often superficial. You can't address a pandemic simply by quarantine. Quarantine might be an important tool for solving the problem, but why use one tool when several might be more efficient? Addressing a pandemic might include changes in social behaviors and in hospital policies. It might include the development of vaccines and travel restrictions. Failing to consider those options — and many others — would be a bit like starting to build a house with only one tool in the toolbox. A hammer is a great tool, but it might not be the first one you'd pick up when you get to the wiring.

We could say more about the characteristics of interdisciplinary learning and will when we consider an example in depth. But before we go further, we want to say that the interdisciplinary approach is not the only way to learn. Some simple subjects really are simple. You can probably learn to add and subtract without getting perspectives from people outside the field of mathematics.

But many topics that human beings find interesting are far from simple. As a rule of thumb, the more complex a topic or problem is, the more a student gains by looking at it from more than one perspective.

One case: Abortion

Let's take one example: abortion.

If you were a student who is interested in medicine as a profession, what would you want to know about the ethical implications of the procedure?

Obviously, you would have to consider the medical perspective — what can and can't be done, what options are available in terms of procedures, what the risks are, what the research suggests about possible outcomes.

But it's also important to consider the legal perspective,

what the law does and does not allow, and how that perspective has changed through the years.

Another perspective is public policy, which considers not only what the law is in this country but what it is in other countries. We can use those questions to ask a further question: What *should* the law be?

What about religious perspectives? They shape the thinking of millions, yet they don't influence others at all. Some religious people follow the teachings of authority figures — the pope, for example. Others do not. What if the patient has no religious qualms about the procedure but her primary-care physician does? Is it appropriate or ethical to express those views in the context of a doctor-patient relationship?

It's also helpful to look at the subject from the perspective of history, to understand how abortion has been practiced at various times in various cultures.

That, in turn, raises the question of how culture affects our perspectives on abortion. For example, women in Texas who are seeking abortions are three times as likely, compared to the national average, to have tried to end their pregnancies on their own. Texas is a socially conservative state, but it shares a border with Mexico, where a drug that induces abortion in early pregnancy is widely available. (Fuentes, L., Baum, S., Keefe-Oates, B. et al., "Texas women's decisions and experiences regarding self-managed abortion." *BMC Women's Health 20*, 6 (2020)),

Such data suggests yet another important perspective. It's important to get information from the fields of demography and biostatistics. No real discussion of policy could go far without support of empirical data.

It's also important to consider the psychological side effects, which can be profound. What if the pregnancy were the result of rape or incest? It's important to include perspectives from the fields of psychology and psychiatry.

There are other ethical considerations, some of which are driven by improvements in technology. For example, advances in imaging technology have allowed us to make great progress in understanding the development of the brain during pregnancy. Many people believe that an early fetus does not have the same status as a viable fetus, much less a newborn. Others disagree strongly, often on religious grounds.

To really understand the complexity of the debate over abortion, you can't just look at it from one perspective — or from two or three. You need to look at the topic from many perspectives to have an adequate understanding.

The reason is that each of these many perspectives influences the others. For example, religious views have influenced and continue to influence our laws. To understand the policy arguments about what good law and good governance would be, you have to have at least some understanding of all the perspectives.

Commitment to the practice of interdisciplinary learning — to getting many diverse perspectives — gives students a more complex picture of the topic under study.

The advantages

What are the advantages of the interdisciplinary approach?

The big one is that it helps people see *connections*. If you are a secular doctor with no religious views, just learning the medical techniques might seem an adequate way to prepare for any ethical questions that might arise. But a physician who has learned something about the religious perspectives on abortion might find it easier to understand the psychological consequences.

There are, of course, other advantages to interdisciplinary learning.

An interdisciplinary approach helps you appreciate alternative perspectives — even if you disagree with those points of view. Being able to see the perspectives of other people is an important brake on the tendency to grasp at oversimplifications. And all too often people and institutions struggle to solve problems because they simply fail to see all the options. Seeing only part of the picture can be disastrous, whether the work you're trying to do is designing an investment portfolio or driving a car on a crowded freeway.

Perhaps the greatest advantage of the interdisciplinary approach is that it increases your analytical options. If you can see a problem from more than one point of view, you can see more than one possible solution.

Again, it's a bit like walking around a sculpture, rather than viewing it from only one angle. Consider what you'd do if you went to see a famous work of architecture, Chartres Cathedral, let's say. If you look at the building from one angle, from the street, you'd see it. But if you move around — go inside and out, up and down, all around the building —you'll see it in new light.

We'd also argue that the interdisciplinary approach is more realistic. Universities are divided into separate disciplines, but the world is not. The view that a single discipline can provide of a big, complex topic is likely to be limited.

Finally, we'd argue that interdisciplinary learning is just interesting. Even in cases where there is no practical or economic payoff for going the extra mile and getting additional perspectives, it's worth doing. That kind of inquiry is interesting, which means it's more fun.

That, incidentally, is true for the teacher as well as the student. Using an interdisciplinary approach allows teachers to learn in fields outside their area of expertise. Having a more interested, engaged teacher is usually good news for the students.

Again, we do not argue that all teaching and all learning should be interdisciplinary. But we do contend that an interdisciplinary approach can often illuminate questions that are simply missed when the inquiry is confined to a single discipline.

Different approach, different methods

Professor William J. Winslade teaches medical and graduate students at the University of Texas Medical Branch in Galveston, but has also taught a course in medical ethics to freshmen at the University of Texas at Austin. Many of those freshmen are considering medical careers.

You can get a grasp of medicine by studying the relevant sciences, including physics, chemistry and biology. But the *practice* of medicine involves dealing with patients, which involves ethical questions. And so students who are thinking about careers in health care are advised to think about the kinds of questions that come with that choice of career. What would you do when a young woman asks for an abortion? What would you do if an older person with an incurable disease asks about euthanasia?

Those are the kinds of questions that don't really lend themselves to true-false or multiple-choice tests.

Writing

Because interdisciplinary learning is aimed at a deeper understanding of a topic, it calls for more than just memorizing some of the relevant material. Students who are going to be facing difficult questions in their professional lives need to be able to reason — and they need to be able to articulate the reasons behind their decisions.

Being able to answer multiple-choice questions is not a good test of that ability. And worse, answering multiple-choice questions doesn't do much to develop your ability to articulate your reasoning.

As the discussion of abortion illustrates, medical ethics is a field that requires a student to look at more than one perspective. The very nature of these inquiries requires a lot of writing.

In his course for freshmen, Winslade tries to examine a student's ability to address complex problems with critical reasoning. He's not trying to assess what each student can memorize.

In looking at the quality of the student's thought, Winslade looks for clarity, coherence and rationality. The way to measure that is to have the students express their thoughts in writing.

The course requires four kinds of writing.

The first is narrative. In the first week, each student must write a short autobiographical narrative. The assignment is not graded but is required. It gets the student used to writing. Later in the course, the student will be asked to discuss advance medical directives with family and friends and write a second narrative that addresses those views, as well as the student's personal views.

The second type of writing is analytical. What does "privacy" mean in the practice of medicine? What does it mean to have "informed consent"? Students write papers analyzing such concepts. In reviewing those papers, Winslade looks for clarity and precision of thought.

The third type of writing is a research paper. Winslade looks for a student's ability to use factual material in arguments that show evidence of analysis and reasoning.

The fourth type of writing is the essay exam. It's an in-class piece of writing, but students are allowed to bring a page of notes. Again, the test focuses on a student's ability to assess and analyze, rather than to memorize.

An approach for all ages

While these examples of

interdisciplinary courses have focused on college and medical students, the interdisciplinary approach can be adopted to every age group.

Elementary students studying the history of the early days of the United States can get a better understanding what life was like by hearing the songs and reading the literature of the era. They can read letters from ordinary people as well as from political leaders. They can get a feel for the technology of the day by seeing how horses were harnessed and how steam engines operated.

During the lecture series on the lifecycle of the brain, elementary and junior high students from Ambassadors Preparatory Academy kept research notebooks that delved into each lecture. The students learned something of neuroscience from neuroscientists and something about early childhood developmental from educators and psychologists. The students consistently asked good questions about a great variety of topics, including minimally conscious states, mapping techniques, and imaging techniques.

The students were exposed to that opportunity not because it was part of a curriculum. They attended because their teachers and parents saw an opportunity to learn.

And, as mentioned before, the lecture series itself was an example of interdisciplinary learning.

Finding one's way about

Ludwig Wittgenstein, arguably the greatest philosopher of the 20th century, used to speak of unsnarling problems — philosophical and otherwise. "The difficulty of philosophy is to find one's way about," he wrote. He was referring to a way of *using* knowledge in a way that would allow a person to see solutions that hadn't been apparent before.

We're arguing for the same kind of reconsideration in our thinking about education. Instead of endless tests of memory, education should be eye-opening, motivating, worthwhile. "Worthwhile" means education is something of value, not in a limited monetary sense, but in the broader sense of the way a person approaches life and the solving of its countless problems.

Education shouldn't be an endless process of memorizing stuff for multiple-choice tests. That's not a good way to teach at the elementary level, and when we give multiple-choice tests to students in medical and law schools, you have to wonder what Americans think education is.

Good doctors don't just rely on memory. Good doctors are able to adjust their style to each individual patient. They realize that you can't treat every patient the same way. They are able to adapt a conversation to each patient.

What we'd hope to get from medical students is not a perfect score on a multiple-choice test but the ability to assess a problem and then help the patient find resources to address it. That kind of flexible thinking is not encouraged by years of memorizing correct answers.

Encouraging independent thinking is good at every level. Younger students should be encouraged to think independently, and they should be encouraged to think independently about their own futures.

The ability to analyze, to research, to write clearly, to think, to solve problems is valuable regardless of what a student decides to do. But it's particularly important for students to use those abilities to think about their own lives, their own interests and their own plans.

A good education and an effective use of the brain produces students that can analyze, research, write, think and solve problems. Do we learn to do any of those things by memorizing stuff?

Chapter 12

Can education be improved?

Rewards and punishments are the lowest form of education. — Chuan Tzu

The philosopher Jacob Needleman, who was educated at Harvard and who taught at San Francisco State University, sometimes talked about teaching philosophy to high school students.

He asked the students to imagine that another person was in the classroom — the wisest person in the world. Each student would be allowed to ask one question. If you were in the class, what would you ask?

"What came back was astonishing to me," Needleman said. "I couldn't understand it at first. About half of the things that came back had little handwriting at the bottom or the sides of the paper in the margin. Questions like, Why do we live? Why do we die? What is the brain for? Questions of the heart. But they were written in the margins as though they were saying, do we really have permission to express these questions? We're not going to be laughed at? It was as though this was something that had been repressed.

"It's what I call *metaphysical repression*. It's in our culture and it's much worse than sexual repression" (Richard Whittaker, "Interview: Jacob Needleman: Art & Philosophy, Oakland, CA 11/21/00," *Works & Conversations*, Sept. 21, 2001).

Shelves of books have been written about the problems facing the American educational system. But Needleman's story gets to the root problem. Children ask profound questions. But adults generally don't do much to encourage children to pursue their own good questions. By the time those children have grown into adolescents, they've learned to control or repress them. One of the challenges of education is to unshackle that sense of wonder.

As Needleman put it: "Modern philosophy often translates the word 'wonder' merely as 'curiosity,' the desire to figure things out, or to intellectually solve problems rather than confronting the depth of these questions, pondering, reflecting, being humbled by them. In this way, philosophy becomes an exercise in meaningless ingenuity (ibid)."

Learning *vs.* "meaningless ingenuity"

What can be said of philosophy, the love of wisdom, can be said of the love of learning. Human beings, by their nature, learn. But this natural activity can be turned into a series of exercises in "meaningless ingenuity."

If you doubt that, consider

Professor Winslade's story about his sixth-grade report on the Panama Canal. It's been decades since Winslade gave that report, but he recalls it vividly. Countless others have similar memories from grade school — moments when the curtain went up, when the love of learning took over. It's a common experience.

Now ask anyone who has had that experience this question: Do you recall a multiple-choice test from that same period? A true-false test?

Most people remember little from their years of education, much less the questions and answers on multiple-choice tests taken decades ago. The usual exams are tests of memory, not learning. There was no joy in memorizing the answers to a multiple-choice test in geography. There was joy in learning about the Panama Canal, exploring an era and a world that were unknown to a young student but that had captured his imagination.

Plato suggested in his *Symposium* that philosophy was born of Eros, a desire so strong that it is was overwhelming. Again, the same thing could be said about learning. You learn when you are interested, when your imagination has been captured, when you *love* the topic you're studying.

And so a grade school student who starts out with an interest about a canal learns something about geography, history, the engineering and transportation systems of an earlier era and the political relations between the

United States and its neighbors.

So why, given what we've learned in the past few decades about the brain, its flexibility and its ability to learn, do we still give multiple-choice tests that essentially test memory? Why do we require teachers to teach to those tests? One plausible answer is that the culture of testing is simply a reflection of our collective anxiety about our educational system.

We are anxious; therefore, we want results. We want measurable results — that preferably can be graded without much effort, ideally by a computer.

Learning is hard to measure. So we measure what we can. The default position is usually a measurement of what's in memory, i.e., what a student can recall. By the time a student gets her degree in English literature, she will have taken countless multiple-choice tests, testing her ability to recall Maya Angelou, James Baldwin and Derek Walcott. One wonders how the masters of literature learned their craft, and whether they ever took a multiple-choice test en route to a mastery of their art.

Learning everywhere

If you look, you'll find advice on learning everywhere. That's because it's important to human beings in all stages of life — to those who have finished their school days as well as to students.

Writing is a species of thinking, and you'd have to look hard to find someone who thought that writing could be improved by memorizing a long list of authors' names, along with the dates and places of birth. Lydia Davis, an American short-story writer and a teacher of writing, starts her essay "Thirty Recommendations for Good Writing Habits" with these points:

1. Take notes regularly. This will sharpen both your powers of observation and your expressive ability. A productive feedback loop is established: Through the habit of taking notes, you will inevitably come to observe more; observing more, you will have more to note down. …

2. Always work (note, write) from your own interest, never from what you think you should be noting, or writing. Trust your own interest. …

3. Be mostly self-taught. There is a great deal to be learned from programs, courses, and teachers. But I suggest working equally hard, throughout your life, at learning new things on your own, from whatever sources seem most useful to you. … (Lydia Davis, *Essays One*, New York: Farrar, Straus and Giroux, 2019)

Most writers will recognize that as sound advice, but isn't that same advice also good for thinking? That is, isn't it good for everyone who has a mind and uses it?

Don't scientists do their best work when they notice things — take note of interesting little things that haven't been explained yet? And don't we all do our best work when we pursue what interests us, rather than what interests an employer or a parent or an academic adviser? And shouldn't we all learn, at a relatively early age, that we can and should teach ourselves? Or, to put it another way, shouldn't we learn that if we're bored, blaming the teacher — or any one else — might be psychologically satisfying, but it won't solve the problem? If you are bored, only you can do something about that.

What about rigor?

Academic disciplines demand rigor. We all have to memorize the multiplication table, the notes on the scale of the treble clef, the proper forms of French and Spanish verbs.

One of the authors chose ancient Greek as his language at university and so spent two years memorizing seemingly endless grammatical forms. He didn't do that out of some perverse desire to punish himself. He did it because he loved Plato, Herodotus and Sophocles.

Learning the endless forms of an ancient language can be drudgery. But, if you're pursuing an interest, it can be a labor of love.

Six suggestions

We've tried, gently, to poke fun at the volume of anxiety about the American educational system. We don't want to add to the library shelves devoted to this topic. Instead, we offer six suggestions for those who want to align educational systems with the way we use our minds.

1. Pursue what you find interesting. Encourage others to do the same. There is really no accounting for personal interest — why one person is fascinated by

biochemistry and another by Renaissance poetry. But if you are interested in something, follow that impulse. Find or discover something new. Learn about it. Teach others.

2. Don't get trapped in academic silos. Avoid specialties for as long as you can. Cross academic disciplines with abandon. It seems that almost all of the really interesting questions cross academic disciplines. One example is the question that led to this book: "How does consciousness arise?"

3. Some universities have programs that are interdisciplinary by design, such as the Plan II Program at the University of Texas at Austin. Plan II ought to be generalized. Interdisciplinary approaches to learning are not just

for the smart kids. And they shouldn't be just for the older kids. It's time to adapt this approach to all levels of education.

4. Early education can introduce kids to things that broaden their vision. Advances in neuroscience have given us clues about how better to use the minds that emerge from the brain, the body and the environment. It's long been clear that languages are most readily acquired at an early age, and the Europeans have been better than Americans at adapting to that reality. Similarly, what we're learning about the emergence of reasoning capacities in adolescents suggests that logic, for example, ought to be taught long before college.

5. We should de-emphasize majors. (Both authors of this book

did that by accident — or against advice — and are unrepentant.)

6. Interdisciplinary learning is not dangerous or seditious. It doesn't replace everything else. It's just a way of making connections, exposing people to things they might be interested in — or might not be. (Learning what you're not really interested in is also a part of a person's education.) But above all, you don't want to miss your interests. An education should help you find your interests. If it hasn't done that, something has gone wrong.

Chapter 13

Using your mind

Long ago, a boy walked out of a library, struggling with an oversized load of books. It was summer vacation. But after a week of baseball, the boy was ready for some reading. The librarian nodded greetings. "It's good to see you are working to improve your mind," he said.

That used to be a mantra, the kind of thing that all librarians and educators said. Everyone has a mind. And, if a mind is something you can use, why not improve it?

The idea was that, if you read a book on George Washington Carver or a book on the biology of a pond, you would learn something about the world that you didn't know before. You will become a slightly better person simply by having a better understanding of the world. You will be better informed, be in a position to make better decisions. It's an idea about human behavior, which is a way of saying it's a moral idea.

•••

An old economics professor, impatient when beginning students got careless with terms such as "price" and "cost," used to tell a story. The cost of the wedding, he would say, is not what the father of the bride pays for the dress, and the cake and the reception. The cost is the difference in the lives that the bride might have led by choosing between suitors. She chose suitor A, instead of B. She will have one kind of life with suitor A, while she would have had another kind of life with suitor B. That *difference* is the cost.

A good novelist could bring that notion to life. Did the bride marry a banker instead of a penniless artist because she wanted a comfortable life? What if the penniless artist sold a painting for $5 million before he was 30 while the banker, who inherited his wealth, quickly squandered it? Would that change the story?

And what of the character of the two suitors? How do you weigh kindness, gentleness, concern, sympathy, warmth?

A novelist could paint a vivid picture of what it is to live one kind of life instead of another.

•••

And so it's time to consider the question of two kinds of lives: one lived with an actively engaged mind, pursuing one's interests and passions, and one lived with a mind left idling.

It's a choice in the same way that the story about the cost of the wedding is about choice. It's a choice between two kinds of life, A and B, each with its costs and rewards.

Socrates famously said that the

unexamined life is not worth living. The original sentence, as reported in Plato, actually says that the unexamined life is not worth living for a human, implying that while cows and horses can live perfectly worthy lives without a lot of self-examination that's just not the case for us.

A book about the brain and the mind must also say something about the moral question of what a mind is good for. This book argues for the old-fashioned view that life lived with an active mind — a life lived learning — is best.

Human beings have brains, and minds emerge from brains (as well as from bodies and the environment). It's the use of that mind that makes us human.

It's an old-fashioned idea, sadly out of step in an age in which most parents spend more time thinking about their children's education in terms of financial security or career opportunities.

We need to change the discussion about our values. Learning is an inherent good that improves the quality of your life, regardless of economic circumstances. Education can be something more than a smart business move.

As we learn more about our minds, we can see clearly that they emerge from our biology. We can also see that individual interests — which vary remarkably — play a critical role in what a person learns.

Knowing that, it seems obvious that learners should be encouraged to explore, to set out early to find their own interests. Decisions about what to do with the rest of your life shouldn't be made until you've given yourself time to explore the options. To find out what you're interested in, you have to expose yourself to a range of possibilities.

It also seems obvious that learning should be lifelong. It's not just for kids. People change jobs and pursue different careers; in retirement, they do different things, indulging hobbies and interests they didn't have time for during their working lives. All these different stages of life call for new skills, new learning. They challenge us to make the most of our minds.

•••

H.D.F. Kitto, who wrote about

the ancient Greeks, said they developed a unique conception of what the human mind was for. Until the dawning of a rowdy democracy in Athens about 2,500 years ago, much of what had passed for intellectual activity in the Western World had involved speculating about fearful gods or chronicling the activities of wealthy rulers.

The Greeks inquired about the world. They asked what made art beautiful. They asked what made water flow, evaporate and freeze. They asked what was the best way to live. They debated those questions — and of course disagreed on the answers. But their answers were *rational;* that is, they gave reasons for their differing views.

One of the standing matters of discussion for the ancient philosophers was the distinction between joy and happiness. Joy comes in moments. Happiness comes over the long haul.

Using your mind to figure things out for yourself at least opens the possibility that you will find things to do that make you happy, satisfied, comforted. To do that you must be engaged — that is, you must find the things that interest you and engage you, not just for the moment, but for a lifetime.

To be a human being involves using your mind to find things that are worth doing and then using it again to do them.

PART III
IMPROVING EDUCATION

Chapter 14

What's wrong with the public education experience?

By SWETA GANTA*

The Declaration of Independence was approved by Congress on July 4, 1776. The mitochondrion is the powerhouse of the cell. The quadratic formula can be used to find x-intercepts of a quadratic curve. If you experienced the American education system, then these facts are most likely embedded in your brain.

Can you explain the motivations behind why the colonists created the Declaration of Independence? Why does the body rely on energy made from the mitochondria?

What is the application for the quadratic equation? Sometime in the past, you might have known the answers to these questions. If your teachers had extra time while lecturing, they could have explained these topics further instead of simply presenting ideas on a lecture slide. Some students are not as fortunate in this respect — they never knew the answers to these questions. American teachers are underpaid, heavily restricted and, in extreme cases of teacher shortages, underqualified. As a result, lessons are straight to the point and are often shallowly explored. In class, information is condensed into digestible units for students to memorize so that more students can be "educated" in a fixed amount of time.

Rote memorization is the enemy of education. Alfred North Whitehead, the author of *The Aims of Education,* warned against "inert ideas" or ideas that live in the mind that aren't used. You can memorize, for example, that China is the number one CO_2 emitter in the world, but that fact alone cannot be applied or used. You need to be able to understand China's culture, infrastructure and businesses to contextualize this fact. Then, it is possible to apply this information if you wanted to think about global warming. In this case, who do you think has a better understanding of the world? The student that memorized that China emits the most CO_2, or the student who knows *why* China emits the most CO_2?

Students are often unmotivated to exercise their cognitive abilities, since their primary methods of evaluation are multiple choice and standardized tests. There is little reason for students to think beyond the learning materials when academic success can easily be attained by memorizing specific pieces of information and reproducing them on exam days. In some cases, exam questions are directly copied from the textbooks or learning materials, so students only need to recognize an answer choice — a low-demand cognitive activity. Learning is a demanding process and requires interest and perseverance, but the current system does not reward students for how well they understand a subject. The current system does not reward learning. Students can put in the bare minimum and still find academic success, so what is the motivation for them to spend more time learning a subject when the ultimate results are the same? Those students who developed a love for learning have done so, not because of the system, but despite it.

The origin of teach to the test

Unmotivated students and underqualified teachers are not the reason why the education system is failing. Instead, misplaced priorities are what led the system to its current state. As a nation, we are concerned with being competitive on a global scale. In terms of education, we want our students to outperform students from other countries, but instead of proposing reforms that would facilitate student learning we opted for shortcuts — completely bypassing learning in favor of results.

Starting in the 1980s, after the nuclear arms race and the space race, the quality of American education was called into question. President Reagan received information that textbook lessons were written to be less challenging, that colleges lowered their admission requirements and that student SAT scores were declining. This prompted Reagan's *Nation at Risk* report, which recommended that schools increase and adhere to standards, introduced standardized testing to assess student progress and prompted colleges to raise their admissions requirements.

The problem of underperformance continued, so the George H.W. Bush and Clinton administrations set goals for students to become top performers in math and science on a global scale. Concern peaked during President George W. Bush's administration and so the No Child Left Behind Act was implemented in 2002. Unfortunately, because of its harsh accountability measures, schools were suddenly expected to meet proficiency standards. Consequently, states either reduced their standards so that more schools would avoid punitive measures, or teachers, out of desperation to keep their jobs, abandoned their curricula and groomed students to pass exams by teaching only the tested competencies and providing sample exam questions. The well-intentioned act then became the driving force for teach-to-the-test methods in schools.

The uses of standardized tests

For anyone that has taken a test, have you ever reflected on why you are taking it? There is an opt-out policy for many states where students are not required to take these exams, but schools are careful to not advertise this option. These tests, while they are said to assess your academic competency, play a more instrumental role in the continuity of your school. Passing rates for standardized tests are used to evaluate teacher and school performance, and if a school does not have at least a 95 percent participation rate it may be subject to sanctions.

Teachers and schools are not the only ones with a stake in taking these tests; students have their own motivations to participate. During transition years, like the 5th and 8th grades, testing scores can determine whether or not students should advance into the next grade. Additionally, high scores provide students entry into elite gifted and talented programs, which help with college preparation and offer opportunities for academic advancement. Standardized tests like AP exams also help students by making college more affordable. Passing scores on AP exams allow students to gain college credit, which allows students to earn their degrees faster.

Standardized tests are also used as a tool for education equality. All students, regardless of background, are expected to meet a minimum standard — otherwise teachers and institutions will face punitive measures. Teachers are then systematically forced to make sure that underperforming students receive an education that will help them pass these exams.

On an international level, standardized tests promote equality because all students are assessed on the same concepts in the same way, and test scores are used to rank countries in terms of ability. In other words, standardized test scores are used to create hierarchies. One criticism of this use of standardized tests is that different countries have different education systems and cultural values that impact the kind of education they receive, so, by assessing all countries equally, we falsely assume that there is one perfect set of education standards and competencies that countries need to master.

Standardized tests not only differentiate between countries but can also be used as a metric for differentiating between people. In the 1920s, high schools and universities were gaining popularity and the competition for college admissions was steadily increasing. Consequently, universities needed a way to distinguish between candidates and so commissioned the College Board to modify an army IQ test and create the SAT. As of 2020, universities, like the University of California and all its satellite institutions, have voted against using standardized test scores in admissions decisions because these tests are coachable — students with more resources and access to prep courses are able to outperform those who do not have the same resources.

If these tests are coachable, then what exactly does a high score mean? Although scores reflect mastery of certain types of skills, if these skills can be reproduced in a short period of time, the main benefit of prep courses, then they might not be the skills that colleges should value. Research also suggests that high school grade point average is a better predictor of college performance than standardized test scores, so these tests are most likely just another hoop for students to jump through and its overall value remains unclear.

Lastly, standardized tests are simple. It would take a longer time to develop a free response test and grade students' individual responses, than it would to create 50 to 100 straightforward multiple-choice questions and run a Scantron through a computer. Standardized tests on a national, state and school level increase efficiency. However, when teachers become accustomed to simple questions, there is less of an incentive to teach difficult material or to thoroughly understand and apply concepts, so students receive a lower quality education for the sake of efficiency.

The toxic testing culture

Standardized tests have become

a crutch for the education system — the only method used to evaluate student progress. In order to get into reputable universities or advance to the next grade level, we are forced to perform well on these exams. For teachers that aren't preparing for national or state-wide assessments, they will often create standardized tests for their classes because it is quick and easy. Whether we like it or not, learning is not quick or easy, and it is definitely not a process that should be made more efficient.

Teaching to the test is a deplorable method of instruction. It is almost disrespectful to boil a discipline down into its smallest parts, sacrificing the interesting and useful, to help students pick the correct letter on exam questions.

We should understand that very few teachers, if any, enter the teaching profession to provide the bare minimum for students. Teachers understand that individualized attention combined with different pedagogical practices can help a student succeed, but the current education system reduces their flexibility in teaching. The strict adherence to state standards, the limited amount of instructional time, the increasing student to teacher ratio, and the systemic importance placed on standardized tests and AP exams prevent teachers from effectively using their pedagogical talents; instead, they become slaves to the system.

Just think about how absurd it is to take a multiple-choice test on writing or reading. If you want to determine if students know how to read or write, ask them a question and allow them to write out their answer, instead of selecting a sentence (written by someone else) that closely resembles their own thoughts. If you want to find out if students can write, ask them to write an essay and then determine if they understand how to use grammar rules and clearly express their thoughts. Reading and writing are practical skills; reducing them to fit multiple-choice test questions prevents students from being able to apply these skills in a naturalistic way.

Teach-to-the-test methods confirm that students have memorized specific aspects of a discipline, but also guarantee that students are limited in their abilities to apply the information in their lives. The bits of information students store in their minds are, as Alfred North Whitehead would call it, inert ideas because they remain unused.

The purpose of education is to develop reasoning skills and build knowledge, but teach-to-the-test methods do not give students the freedom to think creatively about subjects, their potential utility in society or how these concepts can apply to their lives. Students are simply taught to absorb and regurgitate information, not to evaluate and form conclusions. School gives you the tools, but it also is supposed to help you use the tools to improve the way you live your life. What is the purpose of giving people a hammer if they don't know how to use it? What is the purpose of educating students if they don't know how to apply what they have learned?

An alternative approach

Teaching to the test does not provoke students to interact with the material they learn, so whatever information they absorbed has remained unaltered in their minds. These inert ideas help students score well on tests, but do little to help students understand themselves and the world better. What schools need to promote are critical thinking skills.

When we critically think, we are able to evaluate new information, make connections and draw conclusions. By constantly reasoning and evaluating, we are able to supplement, modify and correct our knowledge and beliefs, which helps us grow in our understanding of ourselves and our surroundings. This practice also helps us reinforce old information and apply it in new settings. Critical thinking also helps us defend our beliefs as well as understand other people's points of view because the process of evaluation motivates us to consider multiple options before making an informed decision on what we support. By organizing and analyzing information to reach a logical conclusion, we improve our ability to think through issues, which is a component of problem-solving. Ultimately, critical thinking helps us establish what we know and gives us the tools to determine and correct what we don't.

Developing critical thinking

In order to practice critical thinking, you need to be able to think about what you are thinking about — metacognition. Only from analyzing your thought processes and understanding what you know, can you expand your horizons and gain mental flexibility. Writing is one way to accomplish this. In crafting every sentence, you recall information and you materialize it on paper. You organize your ideas in a logical way to effectively express your beliefs and conclusions. While writing, you need to determine whether what you are saying is grounded in facts or opinions, whether the structure is intelligible and whether your conclusion is the right one to make. Writing integrates the good practices of self-reflection, organization and knowledge application — all of which are skills that, if frequently practiced, will improve the way we integrate and use information. Writing leads to thinking and thinking grows our minds; that is something standardized tests will never do.

---

* *Sweta Ganta is a graduate of the University of Texas at Austin with a degree in Plan II Honors and a minor*

in Health Communications. Transferring from private schools to public schools and participating in a variety of programs across the country, her experience with the American education system has helped shape her values on how to approach learning and increase her awareness of educational disparities. She is now pursuing a career in medicine at the Texas College of Osteopathic Medicine in Fort Worth.

Chapter 15

Interdisciplinary learning in the early years

By JAMES HAUSER*

I believe that elementary schools today resemble industrial factories. The process of education feels akin to an assembly line. It's as if we believe that the minds of children can be processed in the same fashion as goods and products. Subjects and content are mindlessly inputted — covered as quickly as possible with no regard for the uniqueness of each individual person.

The curriculum of each subject/content area is concerned with delivery of instruction that treats children's brains like repositories of knowledge, information, facts, key dates and figures. The more a teacher and school fill up the repository, and the more a child can recall what's in the repository (memorization) the more successful the education is perceived to be.

Comprehension teaching, memorization and even rote learning are not wholly irrelevant, and certainly each can have a place in elementary classrooms. The problem becomes that in an elementary school setting comprehension teaching is constantly used from one subject to the next, hour after hour, day after day, year after year. This repetition stifles the natural curiosity and need for exploration that a child's mind must have in order to grow and develop.

Comprehension teaching is the standard approach in elementary schools, but I have observed that using this approach constantly not only treats students like robots, but it also makes teachers feel like robots.

An interdisciplinary approach

Great teachers love to work with children. In my experience, both as a teacher and administrator, I believe an interdisciplinary approach in primary and elementary education helps teachers and students experience satisfaction that comes from the true success of seeing a child's mind experience the inherent joy of discovery.

Let's start with a few examples for some of the younger students. Notice how in each case the project gets more interesting when other perspectives are added — for example, when an English Language Arts assignment is added to a science project.

Mars, for grades 1 and 2

As an example of interdisciplinary learning in elementary school, consider how first- and second-grade students might learn about Mars.

The planet suggests a science project, and in science class students can learn about telescopes, their usage, and importance. They can watch a video about the Hubble telescope on YouTube. And they can discuss how Mars is visible with a simple telescope.

But Mars is also a good topic for other classes. In English Language Arts, the students can read the book such as *I Am Mars* by Rebecca and James McDonald.

They also can learn about opinion writing. Many people think that we will be able to live on Mars someday. And so the students are asked: "After what you've learned, what do you think? Tell me if you think people will be able to live on Mars. Give at least three reasons why you think we will (or why you think we won't). Use complete sentences and remember to add punctuation."

It's also a good topic for art class. Students can pretend they were chosen to go to Mars. Ask them to draw a picture showing what their house would be like there. And ask them to include an astronaut version of themselves.

Weather for grades 3-5

The weather is a natural topic for math and science classes. While studying collecting, organizing, and displaying data (statistics) in math class, students will learn how to read and create bar graphs, pie charts, histograms, and line graphs. Simultaneously, over the course of several weeks, they will keep track of the weather: temperatures, cloudy vs. sunny, and precipitation. Students will then work on groups to create a display of the weather they have observed based on the skills learned in math class.

P.E. for grades 3-8

How do you measure physical performance — how fast you can run, how high or long you can jump? Ask students to estimate how far they can jump. Students can then make the jump, measure the results and compare it to their estimate.

You can extend the investigation by adding other scientific concepts. Students can learn how to check pulses. They can analyze resting pulse rate, walk a prescribed distance and record the impact on pulse rate. They can run a prescribed distance and analyze pulse rate.

They can then write a report about the differences — an opportunity to summarize the math and science lessons and to develop skills in English Language Arts.

A culture project for grades 2-6+

I witnessed this in Grade 2, but it can be adopted to higher grades. Students identified their family's heritage and researched important topics about culture. Topics included geography, the flag, style of dress, music, dance, festivals and holidays, and food.

It's a way to learn history and geography. It's also a good opportunity to teach basic research skills. As an assignment, students can interview an elder in their family to learn about the traditions and customs.

Students can gather pictures and visual displays from their research. The class and students can then become "world travelers." The students can create mock passports, and the class can travel the world by visiting the different countries outlined in the student projects. The teacher can stamp the passports with each country the students visit.

An extension of this can be to have a Culture Fair outside of school hours to invite family members and the general public. Families can bring traditional foods to serve, and students can display and present their projects. It can be fun for all.

Earth science for grades 4–6+

In science class, students learn about the Earth's axis, revolution and rotation. They also can learn about shadows, types of shadows, the changing size, shapes, directions, and lengths of shadows based seasons of the year and time of day.

On a sunny morning in the schoolyard, students can trace each other's shadows using chalk. They can also trace the shadows of other objects. Later in the day students stand in the same place with a different color chalk and trace their afternoon shadows. Students can also repeat this in different seasons.

To add an English Language Arts perspective, have the students read and discuss the poem "My Shadow" by Robert Louis

Stevenson. Based on their observations in science class and all that they have learned about shadows, students can write their own poems about their shadows.

A better alternative

The interdisciplinary approach provides an alternative to comprehension teaching by developing lifelong skills of critical thinking. Critical thinking skills are antithetical to simply programming students to answer preordained questions. When students are thinking critically, they are coming up with *their own* questions.

Then they read, research, and work to find the answers to questions they have developed on their own. When the answers are discovered there exists a greater satisfaction because students take ownership of the learning process.

The interdisciplinary approach is advantageous because it develops critical thinking skills and instills confidence. Students come to a deeper understanding of individual subjects as they apply what is learned in one subject to the others.

Crossing disciplines, having fun

The "self-contained" setting of the vast majority of elementary school teachers is a natural environment to explore and accomplish an interdisciplinary approach to teaching and learning. There is one teacher creating the unit objectives and essential questions that are thematic, or broad, rather than content specific. The teacher can then create lessons in individual subjects in the same day or over the course of the week that prompt students to find connections to the broader theme.

Some of the more memorable interdisciplinary approaches to teaching and learning that I witnessed occurred in bringing together the disciplines of English Language Arts and American history. As a teacher I remember having fun planning interdisciplinary units, and I remember the empowerment I felt in knowing that I was helping students make important connections across curricula.

For instance, the debate over monuments was very much in the national conversation in 2017. I used an interdisciplinary approach teaching students the fundamentals and principles of persuasive writing in English Language Arts class, while studying the Civil War in history class. As a class we used the current events debate about tearing down Confederate monuments to show the interconnectedness of subjects. The culminating assessment was for students to write an essay that presented an argument in favor of or opposed to the disposal of those statues.

Students were very engaged because they were able to apply the knowledge and facts they had learned in history class with the skills and insights they had learned about writing persuasive essays.

I sent students home with homework assignments to talk at the dinner table and ask their parents, grandparents, and relatives what was their position on the debate.

The interdisciplinary approach developed self-confidence and self-efficacy to be passionate about a current events topic that seemingly everyone in the world was talking about (and still is).

Mixing history and literature

The prescribed curriculum for the study of American history at the beginning of eighth grade in Massachusetts always began with the Civil War era and so I'll describe another interdisciplinary approach involving the subjects of English Language Arts and history.

One summer, in preparing for the school year, I came across a very interesting exchange between Abraham Lincoln and Harriett Beecher Stowe. Apparently, the legend was that when Lincoln met Stowe for the first time he remarked, "So, you're the little woman who started this great war." Whether factual or embellished legend, this quote gave me the idea to teach an interdisciplinary unit — The Civil War and *Uncle Tom's Cabin.*

In American history class, teaching the Civil War begins with the concept of sectionalism and the Missouri Compromise. It can be difficult to impress upon students the importance of the abolitionist movement, and America's growing resentment toward the expansion of slavery with the country's expansion westward. The beating of Charles Sumner, the Supreme Court decision in the Dred Scott case, and the Compromise of 1850 are the ensuing main ideas in American history class.

The Fugitive Slave Act, which was part of this compromise, was a seminal moment in the ultimate breaking of the North and the South. The American history textbook provided all the details, and all the information regarding these important main ideas, but most students were unable to make an emotional connection to the content.

Uncle Tom's Cabin revealed the depravity of slavery and brought out the humanity of the students. I noticed that students who had an affinity for literature, but not for American history, were more engaged in history class and vice versa.

Students were more motivated, discussions in both classes were enhanced, and comprehension in both classes was greater. Ultimately, I believe that teaching the novel created a visceral emotional reaction and connection in the same way that it did to the American public when the book was first published.

Another perspective: music

The following year I added another discipline to this interdisciplinary history-literature unit by inviting collaboration with the music teacher. History tells us that Lincoln took pains to communicate that he was *not* fighting the Civil War to abolish slavery, only to save the union. However, as the war raged on, the messaging began to change, and America's reason for fighting the

Civil War became the end of slavery, i.e. the Emancipation Proclamation.

In teaching this concept in history class I brought the *Battle Hymn of the Republic* into literature class and we analyzed the text as poetry. We analyzed the title, theme, speaker, and tone/mood. This lesson in literature drove home the concept of changing reasons for fighting the war, as discussed in history class.

Then, in music class, the music teacher played a rendition of the song, as well as taught the students the lyrics over two weeks, and they sang it together as a class. This understanding of the content in history class, combined with the emotional connection to *Uncle Tom's Cabin*, and then the significance of *The Battle Hymn of the Republic* created a truly engaging, meaningful, and, I believe, lasting learning experience.

Teaching *Animal Farm*

Another interdisciplinary unit I covered was studying post–World War II, Cold War, and Communism in American history, while studying George Orwell's *Animal Farm* in literature class. This classic allegorical novel is typically reserved for early high school literature classes because it is deemed to be an "easy read." However, in teaching and studying *Animal Farm*. the concept of allegory can be difficult to grasp without the knowledge that the animals (characters) are based on real people and/or tropes of real people.

Having just completed the teaching of World War II over six weeks, I felt I needed to go back in time to the basic concepts of the Bolshevik revolution, the historical persons who led it, the rise of the Soviet Union after World War II, and the impact all of this had on America during the Cold War.

It seemed ambitious, and quite frankly for eighth grade students I thought it could be boring and uninspiring. But the interdisciplinary approach of teaching *Animal Farm* while teaching the aforementioned concepts in history class made both classes more meaningful and enjoyable.

The reading experience of *Animal Farm* was enriched because the literary device of allegory was the objective I sought to achieve in literature, and it “came alive.” I wasn’t teaching allegory by explaining to the students that the animals are people. This is often the mistake that is made in teaching the book, in my opinion. Students lose interest and lack motivation because they’re lacking the historical perspective about what the allegory represents. Through the interdisciplinary approach, students were making the connections to real-life historical persons.

For the sake of teaching the concepts of the Cold War, such as the politics and economics of Communism vs. Democracy, *Animal Farm* enhanced students’ understanding of America’s determination during the Cold

War, and I believe it enhanced their appreciation for capitalism and the democratic values that are an essential objective of American history classes.

Textbook fatigue

In elementary school social studies classes I think that textbook fatigue is a real thing and well documented. A teacher's reliance on textbooks becomes as much an assessment of reading comprehension as it does internalization of the concepts. Chapters are divided into sections, which end in reading comprehension questions.

When it comes to developing students' abilities to write, the textbooks aren't incredibly helpful because their primary function is to assess reading comprehension. In turn, students' abilities and skills of memorization are developed, and their writing expresses a regurgitation of facts rather than emotion.

When the entire curriculum is based on the textbook, and a district's investment is in the textbook, it is only natural that teachers rely on the textbook too much in the delivery of instruction. However, students are not apt to feel emotionally connected to content that exclusively comes from a social studies textbook.

How do you teach Vietnam?

Another interdisciplinary approach I used while teaching middle school social studies and English Language Arts was teaching the Vietnam War while reading the young adult novel, *Fallen Angels* by Walter Dean Myers.

I also sought to use music and video to supplement our use of the textbook to motivate students to appreciate the lasting impact of the Vietnam War. For the book *Fallen Angels,* in English Language Arts class, I was teaching literary elements of theme, setting, plot, point of view, and conflict resolution. I was also teaching literary devices of symbolism, imagery, and figurative language.

It should be noted that Meyer's book was initially challenged as one not to be taught to students because of his descriptive use of sensory details presenting the realities of the battlefields of the Vietnam War. Overall, the experience proved to be further evidence to me that literature is a valuable tool in effectively teaching historical time periods.

The essential objective of studying the Vietnam era in middle school history class is having students understand the causes and effects of the war. First, students must understand that the United States entered the war to stop the spread of communism. This reason to fight a war divided Americans. Second, the American public was further divided by the draft, and an anti-war movement was fueled. The selection of young men for the draft was largely based on socioeconomic status. Those who were able to afford college were not selected for the draft, and the military was primarily made up of people from lower classes.

Third, the effects of the war on American society were another cause of division — the Post Traumatic Stress Disorder faced by soldiers upon returning home, and the loss of faith in government leaders.

In using the book *Fallen Angels* I was better able to emphasize the anti-war movements, protests, and other social issues confronting America during the Vietnam War that was essential for their American history curriculum. In teaching the book, these essential themes in American history class are brought to life for the students and become vivid through the characters of the book.

The protagonist enlists in the army to escape the harsh realities of his life in Harlem, which present him with an uncertain future. He believes he will find opportunity and relief by joining the army and fighting in the Vietnam War. He even believes there will be opportunities for him to become a "hero." But, he realizes that combat is worse than what he left behind at home. The war leaves the protagonist mentally and emotionally scarred.

Writing as you read

Throughout the reading of the book students were tasked with writing their own journal / diary as someone who was in combat with the protagonist and going through the same plot as if it were real life. As students read the book, they had to write journal entries based on the experiences they read about in the book. While they were keeping this journal as a soldier, the emphasis was on the literary elements being used in the book and how they could incorporate them into their own writing.

To enhance understanding of these literary elements and literary devices, I used instructional time in both English Language Arts and history classes to watch *Vietnam in HD* (a History Channel production) via YouTube on the SMART Board. The pictures, video, famous people reading journal entries of real soldiers and nurses, the sounds of helicopters, bombs, jets — all of it was incredibly effective at enhancing the skills I was trying to impart in my teaching of literature while presenting exact circumstances of the war.

The video painted pictures in the

kids' minds and enhanced their imaginations as they read the book. It undoubtedly enhanced their understanding of plot and setting. Even with images presented in the textbook, there is no way that students' emotions would be stirred in the same way. Much like the time period itself, and the way that video brought the war "into the living rooms" of America, the use of video brought the war into our classroom.

I also incorporated the discipline of music during the unit. A couple times per week I would begin class with a song so as to offer the students something different. I would use the songs largely right before American history / social studies class because kids were more excited to read and study

Fallen Angels for a variety of reasons, but because this was an interdisciplinary approach there was hardly a distinction between subjects.

However, generally speaking, American history required more motivational tactics. I used songs that encapsulated the meaningful and lasting concepts of the Vietnam War. Some of the songs we listened to were *Fortunate Son* by Creedence Clearwater Revival, *War What is It Good for,* by Edwin Star, and *Born In The USA,* to name a few.

A culminating project

Combining all of these tools as a means to enhance motivation, aid the principles of comprehension teaching (reading comprehension), and improve students' writing was highly effective.

The culminating project/assessment was for students to write a short story in the genre of historical fiction based upon the journals/diaries they had kept throughout the unit and all of their knowledge of social studies covered in both classes. In their writing students were able to articulate their comprehension of the essential themes of social studies class and the historical perspective of the Vietnam era. Simultaneously, through the connection with literature class they developed a meaningful and lasting understanding of literary elements and devices necessary to enhance their writing and improve their narrative.

Adding a civics lesson

The final interdisciplinary approach that I brought to the classroom as a teacher occurred many years ago when I created a school-wide civics lesson to support troops in combat during Operation Iraqi Freedom and Operation Enduring Freedom in Afghanistan, circa 2005–2007.

While we were studying World War II, my goal was to instruct students on the massive civilian effort on the home front that average people participated in to aid the war effort. I wanted students to understand that the civic duties and responsibilities exercised by an entire generation not fighting the war made a tremendous difference in the efforts of American troops. I was seeking to instill values by having students internalize that the efforts of average citizens became an inspiration and proved to be a difference maker in the quest for victory against America's enemies.

Franklin Delano Roosevelt mobilized the entire population to win the war. There were scrap metal drives, nylon stocking donations for gunpowder bags, victory gardens, food rationing, etc. All of us lived during the global war on terror, and what mind did we pay to the troops? And to the "war effort"?

While studying World War II, our class solicited and collected donations from the entire school community. Our project was for each grade and their families to get acceptable items to donate that would help troops remember comforts of home and let them know they were supported. Items we collected included, Chapstick, sunscreen, DVDs, video games, acceptable snacks, toiletries, and many other items that could be shipped oversees. We organized all of the donations into care packages. The students became the organizers and leaders of it, collecting, counting and boxing all donations.

It was great because it was a service project that built the school community, supported the troops, and enhanced the understanding of how America was able to turn the tide and win World War II.

Most importantly, the interdisciplinary approach of creating a tangible, school-wide service project to connect to history class led to a meaningful and lasting understanding of civics and citizenship, and the role of ordinary citizens in a constitutional democracy.

Exploring the interconnectedness of subjects through an interdisciplinary approach encourages both the teacher and students to engage with subject areas on a deeper level. Using an interdisciplinary approach was something that empowered my creativity as a teacher and fostered meaningful engagement among the students.

As principal, I emphasized interdisciplinary teaching among my faculty. The state may dictate *what we teach,* but as principal I want my talented, gifted, and passionate teachers on the faculty to have the autonomy and courage to decide *how they teach* the prescribed content. My yearly initiative as principal to use the interdisciplinary approach to motivate and engage the interest of students also serves to motivate teachers. The result is meaningful and lasting learning experiences.

Interdisciplinary by design

I strongly believe in the benefits of an interdisciplinary approach in the elementary and middle school grades because I have witnessed the impact it had on students during my time as a classroom teacher. Now, as an elementary school principal, I push my faculty to design interdisciplinary units and lessons. Interdisciplinary instruction is one of my foremost initiatives as an administrator, and it's something that I consistently emphasize throughout the school year.

The long-term goal of the initiative is that the faculty will adjust their instructional practices to the students, rather than trying to adjust the students to fixed instructional practices to adhere to the curriculum of individual subjects.

The school that I lead as principal is a private school serving grades Pre-Kindergarten through Grade 8 and it is on the South Shore of Massachusetts. As principal, I seek to instill a fundamental belief that the importance of our work is not in what we *teach,* but in what we *emphasize.*

With that, the central emphasis of everything we do at these grade levels is predicated on development of a child's ability to understand the nature and function of language. Our fundamental purpose as elementary school educators is to ensure our students can read and interpret critically; that they can write analytically, persuasively, and argumentatively; and that they can speak thoughtfully and respectfully with clarity, diction, and purpose. These skills form the basis for a virtuous and rewarding life.

To foster and develop these skills, the interconnectedness of subjects or an interdisciplinary approach is emphasized. By emphasizing an interdisciplinary approach, we believe we restore meaning and purpose to the educational experience, replacing the stress and anxiety that defines many educational experiences today.

Interdisciplinary units and lessons are essential because as students develop connections between subjects, they develop a deeper appreciation for their educational opportunities; the joy of discovery, and the inherent pursuit of knowledge, truth and wisdom drive the educational experience. This appreciation leads to responsible citizens who contribute to society because their elementary education is more meaningful and lasting.

The faculty embraced the interdisciplinary approach in very creative ways. There were many noteworthy lesson observations in which the teachers employed an interdisciplinary approach with great success. I describe them below in age order offering examples in Pre-Kindergarten, Kindergarten, Grade 3, and Grade 4.

Pre-Kindergarten

Even at the earliest of ages, interdisciplinary teaching can be used to enhance understanding. As an example, at the Pre-Kindergarten level, one of our faculty members incorporated physical education, motor skill development through sign language, and social studies learning by teaching 4-year-old children about Thanksgiving.

Throughout the week students danced the "Turkey Hokey Pokey," which follows the traditional "Hokey Pokey" dance, but it substitutes wings for hands and drumsticks for feet. The dance was about teaching students' physical development and cognitive development. On the physical education side, the dance promotes coordination and self control. On the cognitive development side students are learning the difference between right and left and parts of the body.

It also gets them in the mindset of thinking about turkeys, a traditional symbol of the Thanksgiving holiday for the social studies education component. The class progressed to a song about the pilgrims that involved sign language. The teacher and paraprofessional led the students in a song titled "The Pilgrims Came to America." The song included five words taught in sign language: *America, Thanksgiving, friends, boat,* and *corn*. None of the students had disabilities that *required* the use of sign language. The valuable skills of sign language are inherent, but students at the age of 4 learning sign language is also part of their cognitive development and the development of their motor skills and dexterity.

In addition to these activities, the interdisciplinary approach was also a social studies lesson because the students were learning about the first Thanksgiving.

Kindergarten

In one of our Kindergarten classes, the teacher sought to make interdisciplinary connections in stories that students read. One of the most critical factors for reading comprehension is for students to have a purpose for reading. In this particular lesson plan, the teacher was seeking to make the connection between social studies and science.

The story that the Kindergarten class read was "The Lady with the Lamp: Florence Nightingale." The essential question that students were to keep in their minds as they listened to the story was: What is kindness?

For the social studies curriculum, the teacher helped students summarize and comprehend key details in the story that described Florence Nightingale. The teacher elicited answers and guided students to learn the following information: She was born in 1820.

She grew up in England. Florence's parents wanted her to marry a rich man because that was common for girls in families like hers to do at the time. However, Florence wanted to help people, and she wanted to become a nurse.

The story goes on to make the connection between social studies and science by explaining the work that Florence Nightingale accomplished in field hospitals during the Civil War and Crimean War. Florence recognized that taking care of the health of the soldiers meant showing the army how to keep living quarters clean.

The teacher and story made the interdisciplinary connection by explaining how she taught people the importance of hygiene and cleanliness.

Finally, through this

interdisciplinary connection between social studies and science students learned one way to answer the essential question: What is kindness?

Through the biography of Florence Nightingale, students learned that throughout her life she worked to help others. She kept a journal and wrote about her work so that people could learn from her and they too could care for people who are sick or injured.

Third grade

In third grade, the teacher created a wonderful opportunity for students to read, research, and transform into a historical figure that has made a significant and positive contribution to history. The learning experience began with students selecting a historical figure in Massachusetts and/or American history and finding a biography about that person.

The teacher provided the students with a handout/template, which served as a research guide for the young students to use as they read the book. The template included categories such as birthday, where born, description of family, schools attended, accomplishments in life, what the person was known for, etc. Over several weeks, the students were reading the book at home and during independent reading time in class. At regular intervals the teacher would have due dates and "check-ins" to ensure the progress of students filling in the categories.

She would also take the time to instruct the students on how to properly take notes while they were conducting their research.

From their notes, the students then wrote a rough draft in which they "became the authors" of the person's biography. They would retell the biography in their own words based upon their research. The teacher reviewed and corrected the rough drafts, and then each student took it home to be typed.

Students were also tasked with gathering pictures of their historical figure and decorating a poster board with artwork. They display these pictures, their own artwork, and their typed report, along with important information on a poster board to be displayed in a presentation in class and at a "Wax Museum" in public.

After the written reports, the teacher worked with the students to help them understand the essential elements of public speaking. She reviewed concepts such as self-confidence, eye contact, posture, and the proper volume of one's voice. The teacher also worked with students to write a speech that detailed the highlights and most important aspects of their biography.

Acknowledging the importance of the interdisciplinary approach, the teacher then incorporated acting and role-playing into the learning experience by having the students dress up in costumes to portray the person they researched.

The students come to class in character. In costume, the students recited their speeches and made their presentations as if they were the historical figure themselves.

From this learning experience we observed that in many instances the novelty of dressing in character in many ways "took the edge off" of any trepidation students might have faced about speaking in public. In turn, the students had a meaningful and lasting learning experience in that they learned about the skills necessary to succeed in public speaking.

The culmination of the project came when the teacher set up a "Wax Museum" day in our gymnasium. She invited parents, families, and other grade levels in the school to see the students dressed in character, with their poster board displays behind them. Just as it is when you attend a real wax museum, when one of the visitors in attendance pressed "a button," the students would give their speech and presentation.

Throughout this interdisciplinary approach the third grade students were learning many skills. They learned that biographies are books written about a person's life and accomplishments. Students conducted and learned skills of research to identify important information about their historical figure.

Students were able to use art and be creative in putting together their poster board displays. Students of course learned the lifelong skills of public speaking as the entire learning experience was tied together with the memorable experience of dressing in character and acting out the biography and their research.

Fourth grade

Another instance occurred in fourth grade when the teacher came up with a lesson plan for English Language Arts and math review that she called "Surgery Themed Centers."

All the students were dressed in hospital scrubs, hairnets, facemasks, surgical gloves, and surgical booties — and so was the teacher. There were clear blue tarps all over the classroom, and the entire room had become "The Fourth Grade Hospital."

It was a unique way for the teacher to determine which students were near mastery of the topics during the review. The "Operating Tables" (review centers) included both a review of

English Language Arts class and math class. The centers included: 1) Contraction Surgery Center, 2) Text Feature Diagnosis, 3) Main Idea Skeleton Operation, 4) Write a letter to the surgery board, 5) Pharmacy Arrays, and 6) Using the Distributive property.

For example, at the Contraction Surgery Center, students were given a list of cards with illustrations of children who were "patients." Each patient had the words written on them, and the Band-Aids were the contractions. Students had to work together to match the patient (words) with the Band-Aids (contractions).

At the Text Feature Diagnosis, students used old *Highlights* magazines to find, cut out, and paste the following text features: picture, caption, heading, and map.

Students transitioned from one center to another after 20 minutes; the students had to rotate "Operating Tables." It was an incredible interdisciplinary experience in which learning came to life, and I believe it was a learning experience that will be very lasting for students.

One of the most important aspects of this lesson — and I think the advantage of interdisciplinary learning — is that it was a student-centered lesson. This is to say that the discourse of students with one another was as much a part of the learning experience as the teacher herself.

First, students worked in pairs, and the teacher only needed to facilitate the activities and movements of the students. The lesson was not dominated by the teacher.

Second, the activity and excitement of the students was tangible. At the same time, the students respectfully responded to the teacher's voice. When she spoke, students listened, despite the great energy and excitement they had.

Lastly, all of the centers were reviews of topics that had been covered in previous lessons over the course of the trimester.

Bringing the interdisciplinary approach to the elementary school setting allows students to learn by doing, not simply learn by imitation and memorization. In this way, it creates meaning in the various curricula of different subjects by presenting the interconnectedness of disciplines, just as we experience the world after our time as students in school ends.

** James A. Hauser has worked as a teacher, basketball coach, and principal at the elementary, middle, and high school levels. As a teacher he has taught virtually every traditional core academic subject throughout his career. In his experience designing yearly curriculum maps for various subjects he came to believe in the positive and purposeful impact that teaching with an interdisciplinary approach had on student comprehension and engagement. He earned his master's in adolescent education at St. John's University in 2013 and he is pursuing an M.S. in*

William Winslade and Heber Taylor

Organizational Leadership at Quinnipiac University.

Chapter 16

How can we change the system?

The path of least resistance and least trouble is a mental rut already made. It requires troublesome work to undertake the alteration of old beliefs. Self-conceit often regards it as a sign of weakness to admit that a belief to which we have once committed ourselves is wrong. We get so identified with an idea that it is literally a "pet" notion and we rise to its defense and stop our mental eyes and ears to anything different. Unconscious fears also drive us into purely defensive attitudes that operate like a coat of armor not only to shut out new conceptions but even to prevent us from making a new observation. The cumulative effect of these forces is to shut in the mind, and to create a withdrawal from new intellectual contacts that are needed for learning. — John Dewey, How We Think

By ALEX HARGRODER*

In the years since I began teaching, I've dug further and further into a question about schooling that might seem obvious. Why do we do things this way? Over time, I was able to develop a passable understanding of how our institutions of education have evolved, and even how their purpose has shifted with contemporary currents, and still I'm left to wonder: why this way?

The more I see, the more I realize what we do is determined not by what's best (and, as this book amply shows, we have a clear idea of what would be better). Instead, we often fall back on what we think feels right or lines up best with our memory-shaped expectations. Dewey refers to these tendencies both as a "pet notion" and "a coat of armor." In what follows, I'll talk about how we address this problem with teachers and educational professionals, while also offering a potential map for how we might find a way around these defenses and out of this all-too-familiar rut.

By prioritizing comfort, familiarity, and, perhaps most

damagingly, "efficiency," we have allowed our systems of education to become unintentional replicators of our worst social ills and inequities. What education provides in terms of opportunity is often countered with at least an equal measure of hazard. This has to do both with what we teach and how we structure the environment of learning.

The reality for most students, still, is that schoolwork is rote and disconnected. The school environment is, more often than not, equally hostile, or at least unwelcoming — a place where compliance is enforced and defensive skepticism or apathy can serve as survival mechanisms. This does not match especially well with what we know is best for learning (connection, engagement, reflection), and, perhaps more telling, it does not match with what most people understand the goal of education to be.

In fact, in the professional development work I do with teachers, one of the first things we do is map our vision for what we think student outcomes should be. With almost no exception, these goals involve words like *freedom, agency, power, cooperation,* even *empathy*. Then we ask teachers to contrast this list with what the current system produces: frustration, confinement, inequity.

What's most vexing, though, is trying to pin down a solution to this disconnect. In fact, when asked, most people don't even know where to begin. At play here is what design professionals refer to as "functional fixedness"—the notion that sometimes expertise and familiarity can lead experts to miss obvious, but otherwise unconventional solutions. Because we are accustomed to using tools for their intended function, we can lose sight of alternate (and better) possibilities. This happens when we become so ingrained in the system that we are only able to imagine solutions that exist within it.

Before I go further, I should tell you a bit about myself, perhaps most importantly that I have spent many years in school: as a student, a high school teacher, and as a school administrator. These days I coach teachers.

Given this, it may be no surprise that I mostly liked school when I was a child. Maybe not all the social parts, but the learning, certainly. Given a few years' distance, I now wonder: was it even learning that I liked? Is it possible I had a knack for school in the way that some do for sports? I could listen (convincingly) and make myself small (always) and follow directions (visibly), all exceedingly well, or, more importantly, well enough to please the adults around me, which seemed a higher calling than actually knowing why I was doing what I was told to do. Either way, being seen as a good student came with decent perks, so I kept after it.

It was easy to take these perks as a broader kind of validation. I liked it because it felt good to be seen that way. But what about students who don't fit the mold? Those rebellious students of "bad reputation" who find themselves frequently in trouble for cutting against the grain. What about students who give up, feeling that there isn't a place for them in the system as it is? Whether students internalize (by feeling incapable and unintelligent) or externalize (by acting out) this conflict, my years in the classroom have shown that these students, almost without exception, have unique insight into the cracks in our system. It is from their eyes that we need to train ourselves to see. Their ability to call out unfair or unreasonable things that we've long learned to ignore is key to understanding how we might begin to turn this thing around.

Many students, especially those with budding critical sensibilities, adapt to respond in ways that resist these controlling processes, or even to quietly subvert it, earning the scorn of fastidious adults seeking their own version of control — as maintainers of the system, and too often with arbitrary power and unchecked judgment. However, what may be called a disruption is often instead an opportunity to gain insight into both why and how school isn't working. It stands to reason that those going through the process would have some of the keenest ideas about it. Turns out the view from the bottom is far clearer than from the top.

Think of a very simple act of resistance to these structures of control, like a student saying to a teacher, "This is boring!" to great dramatic effect. First, given what I know about schools, the statement is likely true, and as far as I can tell, telling uncomfortable truths is still fairly transgressive. This bored student is making a public stand, of sorts, perhaps couched in a joke or some bold posturing meant to deflect, but a stand nonetheless against a system that feels meaningless. That's what "boring" indicates here — that this boring thing I'm being told to do doesn't connect to any purpose that I can see.

An outburst of resistance like this, before I learned to see it in the way I'm illustrating, might, in my adult/teacher pose, strike me as disrespectful or insulting, or might even drive me to anger. Really, though, if I were honest, this anger would only be a reflection of my own frustration, one I share with the student (and a bit of guilt about my role in the disconnection). There are few worse feelings than being compelled to prod students through work that doesn't matter — not only is it disingenuous, but it erodes trust, the very thing I seek most from students.

Trust is undermined when students hear from teachers and other adults, "What interests you or makes you curious is less important than what I think." We deny their curiosity our stamp of official approval. Instead we reflexively ask students to obey and follow, and usually without much rationale. Whether we say it explicitly or not, this is the message that our current system is built to convey. And, as teachers, if we reinforce what the system prescribes, we are complicit in shrinking students' worlds and imaginations, not expanding them. This, to me, is malpractice.

In fact, it is trust that lies at the center of what we know as meaningful learning, and thus of change processes. Not only that, but it is what a growing body of research has called a "mediating factor" in countering trauma (cf. Nadine Burke Harris' work on Adverse Childhood Experience Scores). Even more crucially, it is the foundation for all meaningful communities of inquiry. To learn, we reach out into something yet uncharted, which represents a risk, and with this unbounded freedom comes the ever-present possibility of failure. Until we feel safe enough to fail, we will miss the opportunity to pursue mindful transformation.

The question of safety and trust has far-reaching implications. We need to feel safe enough to ask those "troublesome" questions that allow us to step outside of our current set of assumptions and practices. We are often trained to regard these movements as dangerous, heretical even. We offer our teachers a few analogies to illuminate this. The first draws on Thomas Kuhn's idea of a paradigm shift in his landmark *The Structure of Scientific Revolutions*. The idea is that we generally find ourselves in a paradigm, what to us seems "normal," and that so happens to be a system of reference and interrelated parts that, to most people, will remain invisible — a framework that has disappeared into the background.

Kuhn draws on examples like the shift from Ptolemaic (earth-centered) to Copernican (sun-centered) cosmology or the shift from Newtonian fixed-reference physics to Einstein's theory of relativity as examples of ideas that seemed almost inconceivable at their introduction, but, in time, set off revolutions in thought. However, what becomes a revolution may first appear as a crisis. This crisis involves the collapse of the previous system and its concurrent replacement with new and better fitting alternatives.

Especially relevant for our case, though, is what this kind of crisis feels like. It is decidedly uncomfortable, disorienting, even panic-inducing. This is what we're asking people to take on when we talk about transformational and systems-level change. While there's no substitute here for experience, what we can offer, in lieu of instructions and steps, is a map and guideposts.

One useful tool is Meyer and Land's work on threshold concepts, which they describe as disciplinary concepts (for instance, the concept of patriarchy in gender studies, or depreciation in accounting) that, once learned, irrevocably alter a person's understanding of that field in a way that is both troublesome to old assumptions and offers a new way to integrate previously unconnected knowledge. In other words: a new way of seeing. This is what we want for our teachers, and, not coincidentally, for our students by extension.

However, what's most interesting about this is the process that often happens when acquiring these concepts. While Kuhn's paradigm yields to crisis and revolution, then slowly reforms into a new paradigm, the threshold journey begins at the "pre-liminal" stage — our often-uncritical starting point. Into this is introduced some troublesome knowledge — something that throws off the former system of understanding. This leads to a liminal state, one similar to Kuhn's crisis, a rootless and grasping phase where we work to rebuild boundaries and reintegrate our understanding. This is akin to crossing a threshold into a new world, the post-liminal, where we emerge, often after much work, into a new and transformed reality.

I would argue that

interdisciplinarity in education behaves in the same way as a threshold concept. In *The Evolving Self,* Mihaly Cziksentmihalyi captures our current situation by noting that, in the past, students might have easily seen the integrated relevance of their learning. You learn to farm so that you can eat; you learn a tool so you can build. There is no need to reach for a "why." But now, in a world that is mediated by technology and layer-upon-layer of abstraction, we face a new challenge. Our task in education, according to Cziksentmihalyi, is to bring interdependence to the forefront — to teach how life is interconnected, and thus how to navigate and recognize the various systems we both embody and create.

As we tell our teachers, a move toward interdisciplinarity is not merely an instructional strategy, but a process that leads to seeing the world differently. We often kick this off not by introducing the cascade of concepts I just walked you through, but with something simpler. We ask teachers to tell a story about a time when they had undergone a transformation. It could be by way of an experience, a realization, or even learning something new, but nonetheless, we want them to recount in detail a time when they found themselves changed by a new understanding. We ask them to note how it felt then, how it feels now in hindsight, how they changed, and, now, with distance, how they perceive the significance of that change. We've heard answers ranging from tricky math problems to childbirth to the perils of improvisational comedy. Yet all of them follow the same basic pattern.

These stories begin with a disruption or sudden change,

which often quickly gives way to a realization. Coupled with that is sometimes guilt or shame (How could I have ever believed that?). From there we start to build upon that new realization — a bit of learning, a bit of testing, a realization that others might have been in this same position and might know about this too. Soon enough, it becomes a part of us.

In drawing out this pattern, we began to envision a way to recalibrate teachers' way of seeing their roles, as well as curricula and instruction — in a way that honors these fundamentally human (and humane) processes we encounter through our lived experience. There are a number of names for it — dialogic, inquiry-based, cognitive apprenticeship — but they all share the same foundation: there is a meaningful exchange between student interests and teacher expertise. These two forces enter into a third space where they are not in competition, but generative complements to the same pursuit. And in this cooperative pose we find our best entry point into meaningful interdisciplinarity, or what we call a shared engagement with messy, real-world problems, as well as the tools that support this kind of dialogue, like inductive reasoning, embodied learning, and near-constant (and frequently public) reflection.

This alternate paradigm isn't without its complications. It's not always possible to let students go in whatever direction their hearts desire. They may not even know where to head. It's up to us as educators to find points of shared language, tools, or experience that can offer the kinds of permeable boundaries and pathways that enable creativity, versus the potentially terrifying blank page. Interdisciplinarity and honoring students' curiosity doesn't mean that schooling should be a free-for-all, but instead should be responsive to students' needs and what Vélez-Ibáñez and Greenberg (1992) call students' "funds of knowledge"—what complex knowledge and critical capacities students already have by virtue of living in the world, but that may not fit in pre-set curricular containers.

All this is to say: interdisciplinary learning, as a concept, strikes most people as rich and compelling, at least once they know what it is. It may even resemble a more "natural" extension of the types of thematic and inductive thinking we are primed to do when pursuing our established and genuine interests or even on new flights of curiosity. At the same time, I'd get pushback from adults saying it's just not feasible. They couldn't see outside of the system. This I expected. But more startling was the doubt-filled refrain I'd get from a handful of students every year: "This isn't real school — I'm not learning."

It was clear to me that students were learning, but I was fascinated by the idea that it didn't feel that way to them. I saw students gleefully pursuing messy and meaningful questions, drawn from their curiosity and related research. I saw hypotheses emerging and prototypes being tested, yet I still had students insist to me that this wasn't right. So I began to ask: what should it be instead?

The answers I heard were distressing. It became clear that, while students were learning and having fun doing it, they believed they weren't getting the kind of experience they were supposed to be getting. Many of these contrarians were traditionally higher achieving students with good grades and high tests scores. They wanted worksheets and homework and difficult tests. They could not reconcile the experience of what we were doing, even though they were producing richer, better work with more complex thinking, because it didn't have the feel of pressure, anxiety, and stress that the traditional system produces. They saw these experiences as a rite of passage, a hazing of sorts, that would earn them credibility in the adult world as "serious" or responsible.

It's this point, perhaps, that I found most devastating (and, to the same end, might be responsible for why I've devoted myself to this work of change). That students would line up in pursuit of a fairly punishing system because they thought it would give them validation — that very same validation I so eagerly sought when I was young — I simply couldn't imagine wanting to sustain this system.

At the same time, I understand: we revert to what we remember, especially in crisis, or what expectations we've cooked up and harbored based on what we've seen. We want an experience that we share with others — a means of bonding, even. However, this "hidden curriculum" of control and uniformity, what public schooling was originally intended to do in its inception — produce "good citizens" with shared values and beliefs — no longer matches our increasingly global, pluralistic, and interconnected world. This is made clear by nearly every new educational framework or white paper that I read: students need "21st century skills" — the bulk of which involve flexibility, critical adaptability, and the ability to work with others. It is my contention that a fundamentally interdisciplinary approach is the only way to get there.

In order to do that, though, we tell teachers that we have to commit to a few things. First is the mindset that students never come to the table as blank slates, but as rich, critical, in-process people. Their interests and their means of resistance are important and revealing. Second is that we root ourselves in a constant stream of questions and revision, using what we learn and what we find to revise constantly. We are always asking, connecting, and modeling our thinking — so that we can not only see what we think, but we can notice the patterns and systems that we find ourselves in.

Third, is that teachers must be co-inquirers along with students.

Instead of dictating the flow of learning we offer the guardrails of mentorship; we learn with and alongside students. We learn what students are learning and we learn about the students too. We should even learn on our own, but publicly, consciously, as a model for how this can look.

It's no easy task to change a system, especially in education, which is both inextricably political, and, to the extent it sits at the center of both complexities and controversies, quite resistant to change. We must entertain the complexities (organizational structures, content standards, liabilities, procedures) while navigating the controversies (ideologies, what content gets taught, what's "appropriate"). However, if we ground ourselves in a willingness to be transparent, to listen deeply, and to model the kind of trust and care that schooling ought to involve, we can accomplish a great deal, regardless of our circumstances.

It all begins, though, with a bit of productive trouble. It's not the shiny edifice that will offer us the most insight, but instead the cracks, chips, and worn spots. Those give us the best insight into what might come next. Just as urban planners have learned to note and design for "desire paths"—those ruts of convenience that people make when cutting across unpaved areas—so too can we consider taking a step off the prescribed path and into making a connection of our own.

Mapping, making, and reflecting

In my work at MIT's Scheller Teacher Education Program, most of my time involves coaching teachers on how to adopt interdisciplinary project-based learning. For us, this isn't an instructional strategy as much as it's a complicated change process. Our lab (MIT lingo for a working group) deals both in teacher education and the creation of educational games, both of which reflect our strong commitment to situated cognition, which, to keep it short, is the belief that learning happens best when learners are active participants (e.g. doing, making, sharing) and are fully aware of the context (e.g. social, political, physical) of their work.

Much of what I talked about in the first section has to do with the need for change in our educational systems, as well as the change process; what I'll offer here is a brief recap of the mindsets and beliefs that enable an interdisciplinary approach rooted in situated cognition, as well as examples of how this can look when translated to a school setting.

Prerequisites: mindsets and attitudes

First, let's look at a few of those prerequisite mindsets. One commitment that our work relies on is epistemological pluralism. That means we believe that having more perspectives on something (more sources of knowledge) leads to a better understanding of it. Seems simple, right? Not so much, though, if you look at how school is traditionally structured. More often than not, schools teach academic disciplines in ways that don't cultivate thinking or inquiry, but ask instead for memorization and repetition of a singular approach — the "right" answer. While there are, of course, facts and things that are true, the idea that an answer is to be swallowed whole and without explanation (the infamous "because I said so") is both absurd and harmful. Instead, we opt to see those things through multiple lenses, and from this compounded vantage point we are afforded a richer, more fully realized form of knowledge.

The second part of this prerequisite is a bit simpler. We recognize that all students have critical capacities that we can, with time and trust, uncover, support, and grow. These capacities are developed through living in the world; children and adolescents are always learning, even if we may not know how to recognize it when it happens. Contra Rousseau and the generations of educational systems that assume the tabula rasa, the blank slate, we believe students are never empty, but are rich with perspective and insight. As a result, we take a fundamentally dialogic approach to teaching and curriculum. Teacher expertise and bodies of disciplinary knowledge must be in conversation with students' experiences, interests, and contexts. Students know things, and it would be foolish not to begin there.

Finally, we ask teachers to consider the role of emotion in the learning process. More often than not, shame finds a way into our relationship with learning, whether through the specter of failure or the embarrassment of a risk gone sideways. Our job as educators is to root this shame out — not just in practice, but also to push against the structures and systems that support it (harsh grading, inappropriate forms of competition, etc.). This is not only for students, but is something adults should model. Often as a teacher, flexibility and the willingness to admit fault, to revise practices mid-stream, and to simply not know something can seem daunting, especially with the pressures that teaching carries. However, our journey is one of understanding interdependence, not just in the world, but in our classrooms, within our communities, and a cooperative, pluralistic approach is what will offer us a road toward that end. With this in mind, we turn to some of the approaches that can enable strong interdisciplinary learning.

Three interdisciplinary approaches

1. The inquiry environment

Think of inquiry as you would a wave. The wave rises; the wave falls. It takes on structure; it opens back up again. This is the rhythm of an inquiry environment. Structure, space, structure, space. Just as in an art studio, students are shown a technique, have space to practice it, and then get feedback on that practice, so too does the inquiry environment follow this pattern.

One way to think of inquiry is as a system for listening and responding that is rooted in meaningful questions, ideally ones derived from the messiness of life. Being able to root learning in authentic contexts allows for the sometimes abstract skills and techniques, or those smaller parts of a process that may not make sense on their own, to take on a new relevance. When the overarching purpose is clear, coherence is not far behind.

One way we accomplish this is by offering students a protocol to refine their ability to generate questions. Designed by The Right Question Institute, the Question Formulation Technique (QFT) can help students get into the habit of noticing details, seeing contexts, and generating a field of questions that can drive longer-term (and seemingly organic) interdisciplinary inquiry.

The QFT begins with a stimulus of some kind — a photograph, a piece of text, perhaps a demonstration, even a data set — that should be both accessible and complex. Students are given sticky notes and are asked to write as many questions as they can in a brief amount of time (a few minutes at most), one per sticky note. After, we ask students to reread their questions, then to share with a friend. Sometimes, depending on the age-level and complexity of the material, we might allow students a second pass to generate more questions.

Beyond this, we will start to evaluate the questions — whether they are "open" or "closed" (whether or not they have a clear, definite answer) and what sorts of groups they might be sorted into. Because the questions are written on sticky notes, they can be physically moved and rearranged. In this sorting process, we don't prescribe categories, but ask that participants come up with headings for the groups they create. Finally, we ask for the best few questions — those that participants judge to be richest and most important to the topic at hand, or even the ones they're most curious about. From there, we often move to craft an essential or framing question — the kind of macro-question that can guide a unit of inquiry.

Here's an example. One of my colleagues served on a jury just before our last major professional development event for teachers. After her service concluded, and as we hastily prepared our materials, we learned that the case involved a man who had been incarcerated as a juvenile and, upon release, had great difficulty adjusting to life outside the carceral system, and in that crisis committed an act of violence. We used a pair of photographs about the case from *The Boston Globe* (capped with the headline, 'I been in hell all my life') as a stimulus for the QFT demonstration. As a slight modification, after the QFT process we offered a bit more information (the article from *The Globe* in full, which ran at most a page and a half).

Keep in mind, the participants here are teachers, but what they put together is not unlike what we've seen with high schoolers, perhaps just a bit more complex. After the generative rounds (rapid-fire question writing, sorting), we asked these teachers to take their questions and repackage them into a more open form — more like framing questions for a unit. Here's an example of what they generated, all of which reflect the complicated, abuse-riddled childhood of the subject of the story:

What role does society have ultimately in the lives of children?

Does the role of a father really make a difference?

Do educators and other wraparound supports really make a difference?

Does our understanding of mental health change over time?

Is an updated understanding of mental health reflected in treatment in a prison setting?

How does a person's age change their responsibilities and blameworthiness?

How do the perspectives of an imprisoned person, victims/families, prison staff, and policy makers differ?

How did the "tough on crime/drugs" era change the prison system?

What should the role of police be in schools?

What might happen if police were more culturally aware?

How can we use stories, art, history, and data to reform our prison system?

Is there a financial argument to be made that supports education and rehabilitation over incarceration?

Is there a point at which we shouldn't engage education and rehab and simply incarcerate?

How do other countries approach raising children?

How has our society changed our beliefs about raising children?

To what extent should the state (or community via government) be responsible for helping new parents?

What has happened in other countries where there has been radical criminal justice reform?

How much of our justice system (law, legal frames, etc.) is rooted in legacies of racism and discrimination?

How does being in a position of power (e.g. law enforcement, judge, elected official, etc.) affect someone's way of seeing the world? Can we measure this?

From this framing, teachers chose three major questions for this unit:

What has led to the inequity in our criminal justice system?

What is the role of a community in raising children?

What is blame? Who gets to assign it?

These questions are what teachers generated in a very short session and without real editing. This is not meant to be a "right" answer, but an illustration of what sorts of patterns can emerge from a group working together, combining their experience and knowledge, in order to cultivate their curiosity. You'll notice that neither list of questions is bound by a particular discipline, but instead a pursuit of deeper understanding of an entangled, and, as you might notice, quite tragic situation. Socially complex problems like these are especially appealing to students, because they not only offer a window into the systems that underpin our society (that often feel inaccessible to students), but also offer practice with systems thinking, the kind of relational knowledge that leads to insight into interdependence. The questions are not particularly neat and tidy, but compel instead a response that requires boundary crossing, investigation into background information, and a grounding in lived experience in order to make sense of the information presented.

I want to point out a subtle shift I made above. The QFT is a technique that we use with students, but, as you know, my work is primarily with teachers. This is no accident. In fact, as much as possible, we seek to create for teachers the same sort of experience we'd create for students: that of an inquiry environment. Let's look at Tharp and Gallimore's model of "The Teacher's Role in Creating an Inquiry Environment" to illustrate this.

The foundation of inquiry is the question and response process. You'll see this at the center of the diagram. It's a cycle with no clear beginning. Teachers model, question, offer feedback, and instruct. Key to this is a teacher's willingness to model and engage students in an authentic way — as a co-inquirer, rather than a holder of answers. This, in many respects, is the foundation of an inquiry environment: the teacher's example. This is not to discount expertise or preparation, but instead as a reminder that the best inquiry comes from real questions and the risk they involve. If we know the answer already, the question is not of much value. It's what we can't yet see that keeps us coming back for more. And if we aren't willing to try it ourselves, why should anyone follow us?

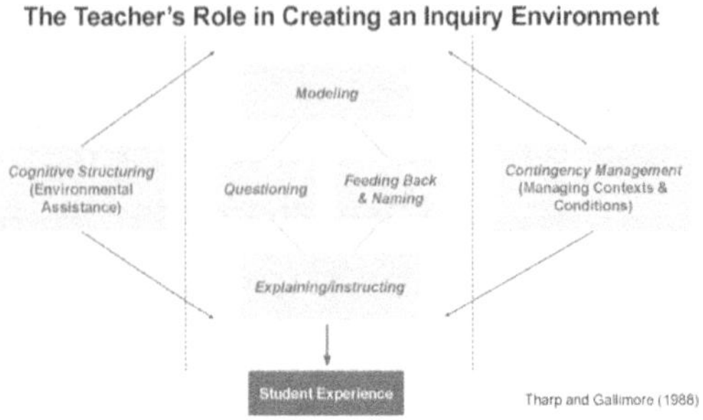

FIGURE: THE TEACHER'S ROLE

However, this is not to neglect the outer two channels of the figure above. While most teachers take quickly to the centrality of the modeling, instructing, question and response cycle, there can be more difficulty with what I'll call the environmental aspects of inquiry. In order to sustain the kinds of questioning I've outlined above, students will need both a significant amount of cognitive support (organizational frameworks, examples of questions at different levels, practice with different perspectives) and contextual support, like the permission to ask questions that might seem odd, or the sense of trust that prevents shame from blocking a student's curiosity. Cultivating this environment is often the most difficult, and, not surprisingly, the least visible aspect of inquiry-based instruction.

Making thinking visible

Speaking of visible things, our commitment to situated cognition rests on one of the oldest forms of human learning: apprenticeship. In case you aren't familiar, apprenticeship works in a few simple steps: the apprentice watches the expert, noticing, perhaps with help, the details of the skills, but always with a sense of what the work is for (repairing a shoe, building a fence, etc.). The apprentice then practices, with supervision and guidance, these same processes, all while building a mental model for the overarching task. In time, the apprentice becomes competent in all of the sub-skills that the craft involves. With more practice comes an even more detailed map, and the ability to improvise, imagine, and understand the contribution of each small part to the greater task. In our understanding of schooling, we take a similar approach through the practice of cognitive apprenticeship.

The foundation of cognitive apprenticeship is also fairly simple. We do what we would with a physical craft, but with ideas. In order to do this, though, we have to make thinking visible. Writing, of course, is the most common way to do this, but there are a variety of ways to broaden writing to include a variety of media and forms of expression, and the willingness to be flexible to these ends. However, most important to this process is the broader frame — the purpose or the ways our work is grounded in the real world, so it's this we turn to first.

In order for students to be effective inquirers, they must first have a sense of their parameters — or, to rely on a geographic metaphor, the lay of the land. What's already there? Another way to refer to this is frontloading or concept-building. When engaging students with a new topic, it is important to offer pathways into that topic — exposure to context and knowledge that students may not have or that may lie dormant in their minds.

As a teacher, I've often challenged myself to build a map of the concepts I hope to dig into with my students. Below you'll find an example of my attempt to map the topic of "juvenile justice" into patterns and clusters of meaning through a broad, but quick (one afternoon) multidisciplinary research session. I did this in advance of our class's reading (I was an English teacher) a number of literary texts involving juvenile incarceration. I wanted to understand how the system worked — how the different causes, responses, and expressions of this concept could be seen together.

FIGURE: JUVENILE JUSTICE

What you see above is not exactly a system, but an array of lenses, both complementary and distinct. My example is probably more complicated than is necessary for good teaching, but I wanted to demonstrate the almost limitless complexity that can be drawn out of a topic, with each sub-topic and detail representing a potential area of study and engagement. Ideally such a sread can allow students to have the opportunity to make choices, but still within boundaries that allow for shared conversations.

There are, of course, much simpler versions of making thinking visible. One example is mind-mapping, which translates to little more than representing a system of relationships between ideas. It might seem daunting to ask students to translate their brains onto a page, but there are much smaller scaffolds (those smaller sub-skills, just like an apprentice) that can make it far easier. For instance, students can, as they read a literary text, record a character's features, whether physical or behavioral, even psychological. This need not be more complicated than a list. However, when this process is repeated a number of times, or is done as a running record, a base of information emerges, which then allows us to find links, ask questions, and make comparisons among the different features or even among different characters. Why did Beneatha reject her wealthy suitor?

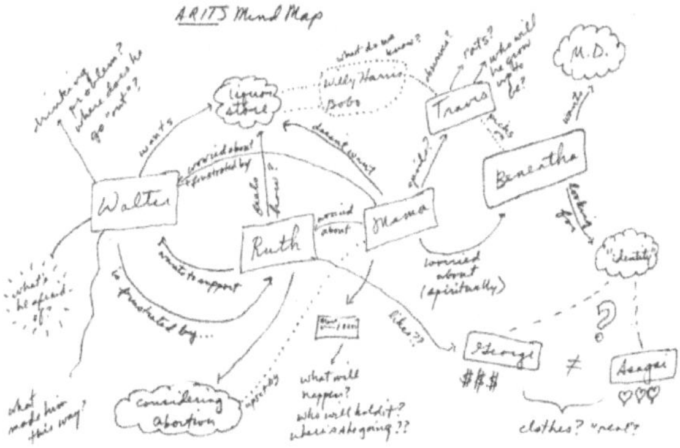

FIGURE: MAP OF *A RAISIN IN THE SUN*

Why did Ruth hide her pregnancy from Walter Lee?

In the map above, made about halfway through a reading of *A*

Raisin in the Sun, you can see the complex relationships between characters, their motivations, and how these might correlate to their development, their position in society, and other social structures (buying a home, housing discrimination, de facto segregation, addiction, gender stereotypes, etc.) that compel interdisciplinary thinking.

Another way to build toward this kind of systematic understanding is through the long-term tracking of thinking by way of reflective journaling. Seeing how our understanding evolves over time can be especially helpful in learning about ourselves as learners. One way to do this is to use what I call a simple "collector / reflector" journal. For students, we often make this digital, using a tool like Google Slides. Students can take a picture of something from the day, whether it's something they found or made or wrote, then annotate it with notes about why it was interesting, questions that linger, or observations that might come in handy later. We keep it simple: one slide per day, just a few minutes. The real magic comes in the repetition. Day after day, while just a quick recording, a complex portrait of a thinker in development, while also reinforcing a habit of mind to stop, notice, reflect, and put words to each new discovery.

With all this visible thinking, you might wonder how we keep track of what learning is happening (i.e. Where are the grades?). Here's where we can loop back to what I shared earlier about systems and change. For that, we have to take a step back and consider the purpose grades serve. In theory, grades should offer feedback on the quality of a task or its product. Often, they fill another role, sometimes to punish what's considered to be inappropriate behavior or to sort students into groups and tracks based on perceived ability. If we were to try to capture only that first part — the evaluation of skills and products — we might have a system that looks more competency-based.

This means that we begin by outlining what Bill Spady calls "life performance" outcomes, those skills and dispositions that underpin or integrate many of our disciplinary tools — things like effective communication, meaningful collaboration, or creative thinking. These holistic measures are meant to capture the process, as well as to illustrate where the learner is on the pathway to expertise.

The figure below includes a few examples of how competencies might be sorted into categories, as well as sub-categories.

There are a number of organizations that have created and adopted systems such as these, and certainly at a greater level of sophistication than I show here.

Interpretive	*Context-Aware*	*Exploratory*	*Connective*	*Reflective*
Understanding & Navigating Systems • Sense-Making & Analysis • Organization & Information Curation • Abstraction, Archetypes, & Symbols	**Understanding Cultural Contexts & Narratives** • Identifying Values & Cultural Elements • Using Context to Situate Thinking • Developing Conceptual Frameworks • Analyzing Narratives & Meaning	**Understanding & Responding to Ambiguity** • Deconstructing & Untangling Complicated Situations & Material • Adapting & Adjusting in Unstable Contexts • Reframing & Evaluating	**Understanding & Navigating Power** • Evaluating with Lens of Equity & Justice • Analyzing Power Relationships • Unpacking Hidden Codes & Structures	**Understanding of Self & Agency** • Interrogating Values & Beliefs • Investigating Strengths, Boundaries & Limits • Discovering & Embracing "Authentic Self"
Using & Manipulating Language(s) • Visual & Linguistic Representation • Evaluation & Persuasion • Adjusting for Audiences • Data Manipulation & Numeracy	**Using Inquiry & Questioning to Expand Knowledge** • Framing & Questioning • Critical Analysis to Drive Inquiry • Applying Multiple Lenses	**Using Creativity & Iteration to Explore & Synthesize** • Constructing & Prototyping • Thinking Hypothetically & Imaginatively • Testing & Experimentation • Using Multiple Perspectives to Synthesize	**Using Empathy & Listening** • Seeking Diverse Perspectives • Listening Actively to Understand • Connecting through Empathy	**Using Reflection & Feedback to Improve** • Setting Goals & Evaluating • Seeking & Using Feedback • Reflecting on Habits & Actions (Both Private & Public)

FIGURE: SORTING BY CATEGORY

But the general principle remains: we learn disciplinary tools in service of these broader, holistic categories that are meant to parallel lived experience, especially that of life in the shifting landscape and rapid technological advancement of the 21st century.

These competencies compel performance and experience — and the closer we get to real-world application, the better. With a bit of planning, what we do in school can be real work for a real audience while still offering the support for development that students need. In terms of interdisciplinarity, there are a number of modes that can capture these performances, including projects, simulations, presentations, designs, events, and any number of real actions that can influence the greater community.

As I mentioned at the outset of this chapter, when asked about what they want for their students, teachers overwhelmingly speak in these kinds of terms, unprompted. If we really want our students to have agency, to be powerful, to be good citizens and people, then we can set these as our desired outcomes, then build an apprenticeship cycle to pursue them. We can map the journey, figure out points to stop, measure, and reflect, and even set out criteria that can tell us if we've succeeded. If we can do this, then we're well on our way to adopting a much more brain- and 21st century-compatible system of education.

Induction & reflection

This last section does not differ a great deal from the two before it. Hopefully, you see the patterns starting to accrue. In fact, that's what this section is about: trying out a sometimes controversial approach, which I call the inductive flip. Bear in mind, my use of the terms inductive and deductive are not meant to hold up to philosophical scrutiny, but instead to illustrate the general shape of things.

Instead of taking a more traditional, deductive approach (start with theory, form hypothesis, find evidence), we might stand to benefit from more frequently, but not exclusively, employing an inductive approach. This means we look at a set of data, perhaps even a messy situation plucked from life, and finding the patterns that it holds, taking a look from a number of different perspectives, then from that generating a working theory that explains the system that holds this data together, whatever it might be.

There are even ways to use an inductive approach to surface students pre-existing knowledge and beliefs. Here's an example based on one of my favorite questions for students: "What is home?" I'd first ask students to write freely about the idea. I'd offer no guidance, no structures, just the imperative to write as much as possible in the time allowed (not terribly long). We would take that writing, which we'd call data, and evaluate it ourselves, looking for things like emotion, words that seem to string together, and the like. Then we'd pool our resources, with me at the board acting as scribe, while students shouted out different words and phrases they associated with home. Occasionally, I'd stop to ask a follow up question, or to have students ask questions of each other. Here's an example of what you might see on my board during this process:

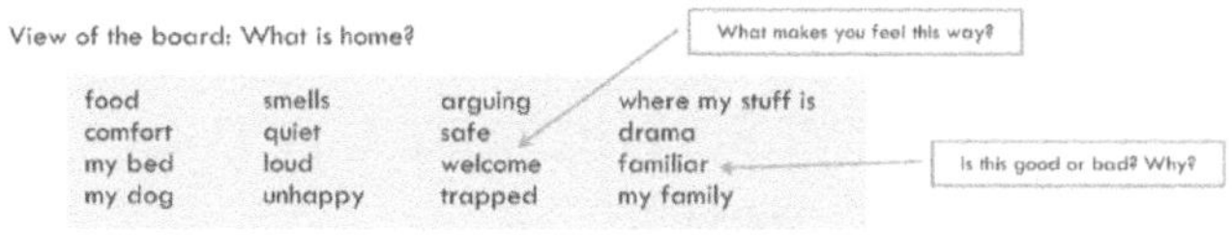

FIGURE: VIEW OF THE BOARD

This is, of course, simplified and cleaned up — often what we'd make on the board was much messier and had notes that indicated when a term was repeated. From here, we could work toward asking more granular questions, some about why the words seem to vary in emotion, perhaps some about whether home is a place or a mindset, and far more than that if we had the time.

I'd use this as a frontloading and concept-building exercise to kick off a unit that dealt with stories of home and growing up, but the general principle holds for any concept or topic. It can be generated from student experience, as my anecdote of home, or from a set of primary sources, or a collection of case files, or even a clip from the news.

The main idea is that, instead of coming in with a pre-conceived notion of what's there, we instead take a moment to see it with fresh eyes, ideally combined with the perspectives of others, which can help us to wedge ourselves out of any mental ruts or functional fixedness due to our prior knowledge in the area. This sort of shift, one that focuses on perspectives, processes, and questioning beliefs runs parallel to several shifts in the way research is being conducted in higher education, including standpoint theory in philosophy, grounded theory in the social sciences, and design-based research in technological fields.

And with that, I want to conclude with what I consider the cornerstone for each of the approaches I outlined above: reflection. Central to all of what I've written is the idea that we are going to be thinking about our thinking. This meta-cognition can be dizzying and daunting, because it can often lead us far from concreteness and real examples. However, it's also what can enable us to grow and evolve as thinkers and learners.

Reflection is a habit of mind that can offer us insight into parts of ourselves we might not notice or remember, while also opening up our thinking to others. By putting our thoughts into language and making them somehow concrete, we magnify our potential and offer a portrait of our emerging understanding, which can not only invite collaboration from peers, but can allow for clear and meaningful feedback from a teacher, which is core to the inquiry process.

In pursuit of an interdisciplinary approach, we may not be able to avoid some of the rigid systems of schooling. We might find ourselves in narrow discipline-based classes and searching for coherence, even for relevance. Given that, we can still employ approaches like these, those that open up the topic at hand, both for us and potential fellow travelers. We can see disciplinary tools, skills, and knowledge as forming a toolbox that helps us to construct a world of meaning. We can commit to grounding our work in a context, and to locating and exposing the systems at work beneath the surface. This is, to me, the essence of interdisciplinarity — the ability to connect, to see not just from one perspective, and to take on questions that matter in pursuit of a life well lived.

---

* *Alex Hargroder is an instructional coach and designer at the Scheller Teacher Education Program at the*

Massachusetts Institute of Technology, where his work primarily involves training teachers in project-based learning, pursuing school change through coaching, and translating transformational learning practices to the high school setting. Prior to this, he taught and served as an instructional coach and administrator in public high schools in Texas and Louisiana.

Acknowledgments

This book grew out of the Robert and Russell Moody Lecture Series, which was cosponsored by the Moody Foundation and The University of Texas Medical Branch's Institute for the Medical Humanities.

Funding for the Institute's Visiting Scholar Program was generously provided by the Sealy Smith Foundation and the National Endowment for the Humanities.

This book was funded in part by the Dieboel Monograph Fund in memory of Donny Rae Dieboel.

Allan Matthews, grants director of the Moody Foundation, helped formulate the idea of the lecture series.

John Paul "Paulie" Gaido, who worked in the background to provide ideas and support.

Dr. Courtney M. Townsend, chair of the Surgery Department at the University of Texas Medical Branch, provided invaluable guidance on the lecture series when he served as interim director of the Institute for the Medical Humanities.

John Zendt, president and chief executive officer of Moody Gardens Hotel, arranged for the lectures to be at the IMAX Theater. The staff of Moody Gardens IMAX Theater helped with arrangements for the lectures, and the crew of Antonio Cisneros and Joel Lockett produced video recordings that are online at https://www.youtube.com/channel/UCRYPnZtw80k42W3O4w3NdlA/videos.

Beverly Claussen and Debra Chambers helped with the funding arrangements. Marcy Sifuentes provided logistical arrangements. Christi Retzer, who was graduate program coordinator for the Institute for the Medical Humanities, contacted the speakers and helped make arrangements for the travel and accommodations. She also helped with the video arrangements.

The Galveston County Daily News published information about the lectures. Many people who attended said they learned of the speakers and their work through the newspaper. Editor Michael A. Smith and Community News

Editor Angela Wilson arranged for the material to be published.

Dr. Patricia Williams, superintendent of Ambassadors Preparatory Academy arranged for elementary and junior-high students from the charter school to attend the lectures. Her students formed a Brain Team, which researched the work of each speaker. Several visited the school. Students visited the laboratory of Dr. Giulio Taglialatela, Lawrence J. Del Papa Distinguished Chair in Neurodegenerative Disease Research at the University of Texas Medical Branch and director of the Mitchell Center for Neurodegenerative Diseases. Dr. Taglialatela kindly arranged for the students to see some of the brains that researchers were studying.

APPENDICIES

Appendix 1

Brief biographies

The organizer and speakers in the Moody Lecture Series

William J. Winslade (Organizer)

James Wade Rockwell Professor of Philosophy in Medicine, Institute for the Medical Humanities and Department of Preventive Medicine and Population Health, The University of Texas Medical Branch. He also is a professor in the Psychiatry Department.

Education

B.A., Monmouth College, Monmouth, Illinois

Ph.D., Philosophy, Northwestern University

J.D., UCLA School of Law

Ph.D., Psychoanalysis, Southern California Psychoanalytic Institute

Biographical highlights

Most of Winslade's career has been spent at UCLA, the University of Texas Medical Branch, the University of Texas at Austin and the University of Houston Law Center. He has taught medical students, graduate students, law students and undergraduates. He has used interdisciplinary teaching in all his courses. His interests include philosophical, legal and psychoanalytic concepts applied to the study of human values and ethics in science, medicine, technology and law. His books include *Confronting Traumatic Brain*

Injury: Devastation, Hope and Healing and *Clinical Ethics: An Approach to Ethical Decisions in Clinical Medicine*. Yale University Press, which published *Confronting Traumatic Brain Injury,* nominated the book for a Pulitzer Prize.

Christof Koch

Chief Scientist and President, the Allen Institute for Brain Science in Seattle

Education

Baccalaureate, Lycée Descartes, Rabat, Morocco

B.S. and M.S., Physics, the University of Tübingen, Germany

Ph.D., Max-Planck Institute for Biological Cybernetics

Biographical highlights

Koch is internationally known

for his exploration of the nature of consciousness and the brain. He oversees a staff of 300 scientists, engineers and technicians at the Allen Institute for Brain Science. He studied and worked with Francis Crick, who won the Nobel Prize for his role in unraveling the structure of DNA and later turned his attention to consciousness. Koch is especially interested in how the brain influences our thoughts, feelings and dreams. He's widely known for his writing, which include *The Quest for Consciousness; Consciousness: Reflections of a Romantic Reductionist*; and *The Feeling of Life Itself: Why Consciousness is Widespread but Can't Be Computed*. From 1987-2013, he taught at the California Institute of Technology in Pasadena.

Paul Bloom

Brooks and Suzanne Ragen Professor of Psychology and Cognitive Science and Chair, Cognitive Science Program, Yale University

Education

B.A., Psychology, McGill

Ph.D., Cognitive Psychology, MIT

Biographical highlights

Bloom conducts research on the cognitive development of children. He's particularly interested in moral psychology, and much of his work has explored how humans develop intuitions about moral responsibility. His books include *Against Empathy: The Case for Rational Compassion; Just Babies: The*

Origins of Good and Evil; and *How Pleasure Works: The New Science of Why We Like What We Like*. Before joining the faculty at Yale, Bloom taught at Brandeis and at the University of Arizona.

Susan H. Landry

Albert & Margaret Alkek Distinguished Chair in Early Childhood and the Michael Matthew Knight Memorial Professor in the Department of Pediatrics at the University of Texas Health Science Center at Houston. She is also the Director and Founder of the Children's Learning Institute at the University of Texas Health Science Center at Houston.

Education

B.A., Music Education, Wagner

College

M.A., Education, University of Maryland

M.S., Communications Disorders, University of Texas Graduate School of Biomedical Sciences, Houston

Ph.D., Applied Developmental Psychology, University of Houston

Biographical highlights

Landry is a developmental psychologist whose research has focused on the biological and environmental influences on children's development from infancy through adolescence. She has studied publicly funded programs to see how they affect children. Her work has appeared in more than 140 peer-reviewed publications. She's the author of *Effective Early Childhood Programs:*

Turning Knowledge into Action.

Silvia Bunge

Professor in the Department of Psychology and the Helen Wills Neuroscience Institute at the University of California at Berkeley

Education

B.S., Biology, Yale College

Ph.D., Neuroscience, Stanford University

Postdoctoral fellowship, Massachusetts Institute of Technology

Biographical highlights

Bunge directs the Building Blocks of Cognition Laboratory, an interdisciplinary research group that investigates developmental changes and neural plasticity in children using tools and insights from neuroscience, developmental psychology and education research. The lab seeks to better understand environmental influences on the development of the brain and on cognition. Bunge's research interests include the emergence of reasoning ability across childhood and adolescence. The lab also seeks to better understand how the brain is shaped by experience — that is, how a child's home environment and schooling and other activities affect the structure and function of the brain.

Walter Glannon

Professor of Philosophy, University of Calgary

Education

B.A., Philosophy and Spanish

literature, Duke

M.A., Spanish Literature, Johns Hopkins

Ph.D., Spanish Literature, Johns Hopkins

M.A., Philosophy, Yale

Ph.D., Philosophy, Yale

Biographical highlights

Glannon has written 10 books, including *Psychiatric Neuroethics: Studies in Research and Practice; Brain, Body, and Mind; Neuroethics with a Human Face*; *Bioethics and the Brain*; and *Biomedical Ethics,* all published by Oxford University Press. His book *The Neuroethics of Memory* was published by Cambridge University Press. He previously taught at the University of British Columbia, McGill, Smith College and was Canada Research Chair in Medical Bioethics and Ethical Theory at the University of

Calgary.

Robert A. Duke

Marlene and Morton Meyerson Centennial Professor in Music and Human Learning; Director of the Center for Music Learning, Sarah and Ernest Butler School of Music; Clinical Professor, Dell Medical School, The University of Texas at Austin

Education

B.M.E., Florida State

M.M., Percussion Performance, Florida State

Ph.D., Music Education, Florida State

Academic appointments

Duke, a former studio musician, taught music in public schools and has worked with children at-risk through the juvenile justice system. He has written behavior analysis software. His books include *Intelligent Music Teaching: Essays on the Core Principles of Effective Instruction* and, with Jim Byo, *The Habits of Musicianship*.

Arthur B. Markman

Annabel Irion Worsham Centennial Professor of Psychology and Marketing, University of Texas at Austin

Education

B.S., Cognitive Science, Brown

M.A., Psychology, Illinois

Ph.D., Psychology, Illinois

Biographical highlights

Markman's research interests include work on how people see things to be similar and how they make comparisons, which led to work in category-learning and decision-making. He was founding director of a program called the Human Dimensions of Organizations and has been a blogger for sites such as Psychology Today and Fast Company. His books include *Smart Thinking: Three Essential Keys to Solve Problems, Innovate, and Get Things Done* and *Smart Change: Five Tools to Create New and Sustainable Habits in Yourself and Others*. He is executive director of the IC2 Institute, an organization that focuses on innovation and entrepreneurship.

Duke & Markman: Joint highlights

Duke and Markman are hosts of

an NPR radio show and podcast "Two Guys on Your Head." The radio show has been described as similar to "Click and Clack" — except that it's about minds instead of cars. The show is an effort to bring a little science of psychology into people's week. While the show is about science, it's heavy on humor and focuses on the baffling aspects of human behavior. A collection of highlights from the program is the basis of the book *Brain Briefs*. The show also has some musical notes since both professors are also musicians.

Nicholas D. Schiff

Jerold B. Katz Professor of Neurology and Neuroscience, Weill Medical College of Cornell University, and Director of the

Laboratory of Cognitive Neuromodulation at Weill Cornell

Education

B.A., Philosophy, Stanford

M.D., Cornell University Medical College

Biographical highlights

Schiff is a neurologist and neuroscientist whose work has been recognized by *Time Magazine* and *Rolling Stone*. He was involved in a highly publicized case involving a man who was largely unresponsive after being severely beaten in a mugging. Schiff developed a unique brain pace-making strategy and led the team of investigators who treated the man with deep brain stimulation — a technique that involves implanting electrodes in the brain. The treatment was successful, and the patient's mother spoke of her son coming back. As both a physician and scientist, Schiff is interested in the neurological disorders of consciousness.

Joseph J. Fins

E. William Davis, Jr., M.D. Professor of Medical Ethics and Professor of Medicine, Weill Cornell Medical College

Education

B.A., Wesleyan University

M.D., Cornell University Medical College

Biographical highlights

Dr. Joseph J. Fins teaches at Weill Cornell Medical College and Yale Law School and co-directs CASBI, the Consortium for the

Advanced Study of Brain Injury. He is the author of *Rights Come to Mind: Brain Injury, Ethics and the Struggle for Consciousness*, Cambridge University Press (2005). In 2019, the American College of Physicians named him the winner of the Nicholas B. Davis Memorial Scholar Award, which recognizes "outstanding contributions to humanism in medicine" as well as "scholarly activities in history, literature, philosophy, and ethics."

John Paul "Paulie" Gaido

Lawyer, businessman and medical humanist

Education

B.A., University of Texas at Austin

J.D., University of Texas School

of Law

Th.M., St. Thomas University Graduate School of Theology at Saint Mary's Seminary

Ph.D., University of Texas Medical Branch

Biographical highlights

Gaido is known widely as a restaurateur; the family restaurant, Gaido's, is arguably the most famous seafood restaurant on the Texas coast. But he is also a lawyer and medical humanist who believes in the value of advance directives as a way to exercise your autonomy, especially as you age and face the prospect of dementia or other disabling medical conditions. His writing, including the book *Mapping Your Own Alzheimer's Journey*, has focused on the moral obligation to make decisions about care at the end of life. He also wrote, with his wife, Mary Kay, the book *Entertaining at Home with Fresh American Seafood and Exceptional American Wine*.

Daniel Wikler

Mary B. Saltonstall Professor of Ethics and Population Health, Harvard School of Public Health

Education

Ph.D., UCLA

Biographical highlights

Wikler was the staff philosopher for the President's Commission on Ethical Problems in Medicine under Presidents Carter and Reagan and was an author of the section of the commission's report that resulted in the redefinition of death in nearly all U.S. states and much of Western Europe. He has been interested in bioethics both as an academic and in public service. He was the first staff ethicist for the World Health Organization and remained a consultant to several programs, most recently its consultative group on the ethical dimensions of universal health coverage. His research interests include ethical issues in population and international health. Topics include the allocation of health resources, health research involving human subjects, and ethical dilemmas arising in public health practice.

David Van Essen

Alumni Endowed Professor of Neurobiology, Washington University in St. Louis

Education

B.S., Chemistry, California Institute of Technology

Ph.D., Neurobiology, Harvard Medical School

Biographical highlights

Van Essen has described himself as "a cortical cartographer," a person who maps the brain, particularly the cortex. He is a principal investigator with the Human Connectome Project, which is based at Washington University in St. Louis. In 2016, the Human Connectome Project published a map of the brain that made worldwide news. The map details 180 regions of the brain — 83 that were known and 97 that hadn't been described. Van Essen started his career by developing flat-map techniques — turning a three dimensional object, the brain, into a two dimensional representation, a map — with pencil and tracing paper. Computerizing this process took almost two decades.

Norman Doidge

Assistant Professor of Psychiatry, Faculty of Medicine, University of Toronto; research faculty and lecturer, Center for Psychoanalytic Training and Research, Columbia University, New York

Education

B.A., University of Toronto

M.D., University of Toronto

Biographical highlights

Doidge, an internationally renowned psychiatrist and psychoanalyst, has written two best-selling books: *The Brain that Changes Itself* and *The Brain's Way of Healing*. Both reflect

Doidge's conviction that neuroplasticity — the brain's capacity to change — has been the revolutionary idea in understanding the human brain. His work was featured in the PBS special "The Brain Fitness Program." He is also a poet, having won the E.J. Pratt Prize for Poetry at age 19. At the University of Toronto, he studied classics and philosophy. In addition to teaching psychiatry, he has also taught courses in philosophy, political science and law.

Heber Taylor

Heber Taylor is a retired newspaper editor. His interests include philosophy, the natural sciences and the ancient Greeks.

Appendix 2

An essay on interdisciplinary learning

By William J. Winslade

I

In this reflective essay I will review my personal education, including advanced academic and professional education. I will describe my very early preschool experience through elementary, junior high, high school, college, graduate and professional school. I will then describe my own style of teaching and learning. I will try to show why interdisciplinary education provides a far better approach to education at all levels.

My basic claim is that interdisciplinary learning encourages and invites analytical, as well as synthetic, thinking, reasoning and writing.

When I was about 20 months old back in 1942, I fell from a second-story porch. I landed on my head. The mid-left back of my skull was fractured, a compound-complex fracture. When my grandmother, who had only briefly left me on the balcony, saw me bleeding and semiconscious on the concrete below, she rushed down, picked me up and carried me to the nearest hospital. When we arrived the question was not "What happened?" but "Can you pay?" Fortunately, the hospital workers believed my grandmother, and I was seen by a Viennese primary-care doctor who removed the bone fragments. He told my family he had no idea whether I would live or die. (For a more complete discussion of this incident, see my *Confronting Traumatic Brain Injury*.)

After five weeks in a hospital bed, I had fortunately recovered enough to be discharged. I had remained conscious after the fall and in the hospital. I was under the watchful eyes of three sets of private-duty nurses 24 hours a day. The only medication I was given was a drug to keep me drowsy and quiet. It was cherry-flavored; I still don't like cherry-flavored liquids — but I make an exception for cherry pie.

The only thing my family told me about my recovery was that after five weeks in a bed, I had to relearn how to walk. I also know that I had many visitors during my hospitalization.

Undoubtedly, my family was worried about brain damage — as was Dr. Bernauer. In the hospital and at home, many people — parents, aunts and uncles, grandparents — all spent time talking to me. No doubt they worried that the traumatic brain injury might cause me to be unable to learn. Many years later, a psychologist told me that the frequent verbal stimulation undoubtedly contributed to my recovery and perhaps even enhanced my early learning. I know that they had taught me to read and write by the time I was 4. I have seen and read some of my earliest written products.

When I was 5 years old, my

parents and I moved from Chicago to Carlinville, a town of 5,000 in south-central Illinois. My father initiated his veterinary practice, while I explored the neighborhood. There were no preschool or kindergarten options. When I entered first grade in 1946, I could read and write — though not very well. But I recall being bored and restless in my first-grade class. Perhaps that partly explains my tendency to disrupt class, aggravate the old teacher and have chats with the principal.

Luckily for me my second-grade teacher was wonderful — warm, engaging, encouraging and effective as a teacher. I loved her and the class. I learned a great deal and remained in contact with her throughout my childhood and adolescence. I even played golf with her husband when I was a teenager.

The only other exception to an uninspired and deadly elementary education was Mr. Reynolds, my sixth-grade teacher. Unlike all my other grade school and nearly all my high school teachers, Mr. Reynolds challenged me and my classmates not only to read but also to write a research paper. I still remember spending many hours doing research and writing a report on the Panama Canal.

My early education was otherwise boring and uneventful except for my lively and stimulating seventh grade science teacher. My high school education was also dull and boring, and testing was almost exclusively true-or-false and multiple choice.

My high school experience was very satisfying but it had little to do with academics. I was fortunate to have a music teacher who made the high school band enjoyable, who encouraged us to compete in music contests, to perform in public and to play in the city band in the summer. I played an alto saxophone and by 1958-59 I had organized a small rock-and-roll band outside of the high school. We had a drummer, a piano player, a singer-guitarist and a saxophone. We may not have been great musicians, but we were good enough to play at dances and make decent money.

The other things that made high school tolerable were sports. I played basketball and ran on the track team. Because of my childhood brain injury, I did not like football very much — and thought it was dangerous. I played in the marching band. I later learned that the football coach tried to convince my father to encourage me to play football. But I said, "Why would I want to go bang my head wearing a helmet against other nuts?" I used to say that I did not want another hole in my head.

My high school educational experience was very much like my elementary and junior high experience. Very little homework, only true-false or multiple choice exams, hardly any written work and very little learning. I clearly recall asking my math teachers about the history and theories of mathematics but never got any helpful response. Only later in college when I took a logic course did I discover the fascinating literature on the history of mathematics and logic that I read on my own.

I write about my retrospective negative feelings about my early education because my curiosity about learning was not stimulated until the second semester of college when I took my first philosophy class. My philosophy professor was fascinating, although at first I was puzzled about what he was teaching. I eventually realized that he was provoking the students not to memorize information for tests but to explore ideas, to think about puzzling questions and to read about how we think, reason, assess evidence and form our beliefs. From that point forward, I was inspired to ask questions about knowledge, evidence, arguments and the underlying assumptions about law, ethics, religion and the social order. I became obsessed — in a good way — with philosophical issues that had never been presented to me nor had occurred to me.

And, most importantly, in my philosophy class I was asked to write research papers and answer essay questions on exams. I began to realize that there were so many issues I had never considered, much less researched. My attitude toward education radically changed. Similarly, I was introduced to economic theory and public policies by my two economics professors, one a theoretical economist and the other a pragmatic lawyer.

Philosophy and economics

taught me to do research, to read carefully, to write essays and research papers. That was the end of true-false and multiple choice exams. And in my subsequent career as a university professor I have never tested my students except by essay questions and research papers.

Since my B.A. in college, I have gone on to obtain a Ph.D. in philosophy, a law degree and a Ph.D. in psychoanalysis. I am happy to tell you that all my graduate and professional education has been based on analytic and / or research writing. My interdisciplinary learning has helped me to see connections and understand philosophical theories, legal rules and policies and overt and hidden emotional issues that pervade everyday life, the academic world and personal concerns.

I offer this lengthy personal experience in support of my criticism of much of what is called the educational process. It has become clear to me that teaching to the test and failing to encourage students to learn to think, reflect, analyze, do research and write essays and research papers undermines intellectual curiosity, development and critical thinking. Teaching to the test, which has become pervasive even in much of professional education in law, medicine and other professions, undermines both learning and intellectual reflection.

In this book we advocate interdisciplinary learning and both reflective and research writing as an alternative to and an antidote against teaching to the test.

I have discussed in Chapter 11 how I teach medical ethics to first-year, first-semester students at the University of Texas at Austin. I also teach graduate students in medical humanities at the University of Texas Medical Branch. In all my classes — undergraduate, graduate and professional legal education — I use only essay examinations and research papers as a means for evaluating students' knowledge, analytical and research abilities. But I think we need to go further.

I will sketch an argument for interdisciplinary learning. I strive to do this in all courses that I teach.

II

The primary goal of interdisciplinary learning is to discover, group and utilize obvious and subtle connections among and between different intellectual perspectives. For example, I teach a course on law, ethics and health care. End-of-life issues typically involve personal choices, legal rules, ethical values and health policies. When I teach about topics such as the legal, ethical and psychological aspects of Medicaid physicians' options and obligations regarding diagnosis and treatment of a patient, legal rules, ethical standards and psychological impact are all relevant considerations.

Physicians and other health professionals have duties that go beyond accurate and reliable diagnosis and treatment. They have legal duties, ethical considerations, standards of practice, emotional considerations, economic factors, religious and family issues, psychological issues, and emotional and mental health issues that are all relevant to a particular medical encounter.

The practice of medicine and other types of health care typically requires interdisciplinary learning for appropriate professional care.

Attorneys also must be sensitive to psychological, economic, religious and other forces in representing clients. Effective professional practitioners in other areas have similar obligations and skills.

Teachers at all levels have similar

responsibilities to understand this type of learning, to appreciate how it is related not only to learning but also to the lives and relationships of their students. Teachers must understand how psychological issues enter into personal and interpersonal interactions.

Teachers of professionals, as well as the professionals themselves, live in a world with inter-professional and interdisciplinary issues whether they realize it or not. An effective practitioner of any profession must be alert to the connections — the interdisciplinary connections — that are germane to their own work as well as to the explanations that they give to their clients.

Likewise, an effective teacher — of virtually any academic or professional practice — should be attuned to the ways in which their area of practice and expertise is related to other areas of learning and practice. A concert pianist must not only be able to read and play music, but also have command of the instrument, understand how an audience might react, and be able to deal with the music business, the emotional implications and impact of music, etc.

Interdisciplinary learning is a pervasive background for so many things we do and care about. For that reason, learning itself calls for interdisciplinary learning and understanding to enhance and carry out complex human activity.

What I am arguing is that interdisciplinary learning is essential to many human practices and undertakings. To take another example, consider the case of veterinarians — what my Dad needed to know. A good veterinarian must understand:

- Common animal diseases.
- Current scientific knowledge.
- Attitudes and disposition of farmers.
- Animal emotions.
- Drug policies.
- Treatment options.
- Legal rules.
- Best practices.
- Experimental options.
- Cultural attitudes, such as the attitudes of owners toward life-sustaining treatment or euthanasia for sick or diseased animals.
- Economic policies.

Appendix 3
Media: Op-ed pieces and Q&As from the lecture series

The following pieces appeared before each of the lectures in The Galveston County Daily News *and are republished with the permission of the newspaper.*

The Robert and Russell Moody Lecture Series

1. Christof Koch — Oct. 5, 2016
2. Paul Bloom — Nov. 3, 2016
3. Susan Landry — March 7, 2017
4. Silvia Bunge — Sept. 13, 2017
5. Robert Duke & Arthur Markman, Oct. 25, 2017
6. Walter Glannon, Feb. 7, 2018
7. Nicholas Schiff, April 4, 2018
8. Daniel Wikler, July 12, 2018

9. John Paul "Paulie" Gaido, Sept. 26, 2018
10. Joseph Fins, Oct. 24, 2018
11. David Van Essen, Feb. 6, 2019
12. Norman Doidge, March 6, 2019

1. Christof Koch — Oct. 5, 2016

Lectures will explore mysteries of the brain

By William J. Winslade

If you are reading this sentence, then you are conscious. In fact, we are conscious unless we are in a deep sleep, coma, anesthetized or suffering from a medical condition such as a stroke, brain injury or another condition that disrupts our normal brain function. When we are conscious we can think, talk, make plans and engage in all kinds of human activities. But what is consciousness?

We know that our brains are essential to our conscious life, that our bodies and the environment influence and impact our consciousness.

Brain scientists and researchers have studied living and dead brains to try to understand how brains function. But a full and deep understanding of consciousness has proven to be challenging and elusive. But don't we all wonder how the brain works and wouldn't we like to know more?

Christof Koch, president and chief scientific officer at the Allen

Institute for Brain Science in Seattle, is an internationally acclaimed brain scientist who has devoted many years of his academic and scientific career to the brain and explaining it to us in language we can understand. His books *The Quest for Consciousness* and *Consciousness: Reflections of a Romantic Reductionist* are two of his major works.

In addition he has written many articles. Here is a sample of the titles of some of his intriguing articles: "Does Size Matter — for Brains?," "Is Consciousness Unconscious?," "What Do Babies Really Know?," and "Safely Switching Consciousness Off and On." These and other of his articles help us to unravel the mystery of consciousness.

Koch will deliver the inaugural Robert and Russell Moody Lecture entitled, "Consciousness and the Birth, Life and Death of the Brain" on Oct. 5 at Moody Gardens IMAX Theatre at 6 p.m. This lecture is free and open to the public.

This is the first in a series of public lectures on the life cycle of the human brain and human behavior. The next lecture will be given on Nov. 2 by Paul Bloom from the Yale University Psychology Department and its Infant Cognition Center. Bloom is a distinguished scholar who has revolutionized how to think about the minds and behavior and sense of morality in pre-verbal children. His recent book is titled, *Just Babies, The Origins of Good and Evil*.

This lecture series is cosponsored by the Moody Foundation and The

University of Texas Medical Branch's Institute for the Medical Humanities.

Exploring the development and deterioration of the brain is important for everyone to know, especially students interested in science and medicine.

(Published as "Lectures will explore mysteries of the brain," The Galveston County Daily News, *Sept. 24, 2016)*

2. Paul Bloom — Nov. 3, 2016

What's going on in those little heads?

By William J. Winslade

Do you wonder what goes on in

the minds of preverbal toddlers? Do they only think about food and toys and naps?

Paul Bloom, an innovative developmental psychologist from Yale University, believes that toddlers have innate moral intuitions. In a *New York Times* article he says: "a team of researchers watched a one year old toddler boy take justice into his own hands. The boy had just seen a puppet show in which the puppet played with a ball while interacting with two other puppets. The center puppet would slide the ball to the puppet on the right, who would pass it back. And the center puppet would slide the ball to the puppet on the left … who would run away with it. Then the two puppets on the ends were brought down from the stage and set before the toddler. Each was placed next to a pile of treats. At this point the toddler was asked to take a treat away from one puppet. Like most children in this situation, the boy took from the pile of the 'naughty' one but this punishment was not enough — he then leaned over and smacked the puppet in the head."

One of the questions I have for Professor Bloom is whether, in addition to the innate moral intuition on punishment, the toddler also displayed an impulse to violence when he smacked the naughty puppet.

In thinking about this study, I also wonder whether some toddlers might respond very differently to the puppet show. A toddler might simply be upset and not want to do anything after the game was disrupted. Or the toddler might want to just continue the game with the good puppet. Still other toddlers might want to reward the good puppet and then resume the game.

I am sure that Professor Bloom can enlighten us further about the behavior of toddlers. He will lecture about "Just Babies: The Origins of Good and Evil," based on his book by the same title.

It is a rare opportunity for the Galveston community to hear about the fascinating research conducted at the Yale Infant Cognition Center. In a world filled with violence from shootings to assaults to wars, one hopes that we can find a way to educate children as well as adults to find effective alternative to controversies and moral disputes that are less destructive than violence.

It will not surprise me if people who attend the lecture will also have questions about the origins and evolution of morality that Bloom can answer. His research and insights are presented in language that anyone can understand and also appreciate.

(Published as "What's going on in those little heads?" Oct. 28, 2016, The Galveston County Daily News*)*

+++

Q&A with Paul Bloom

Q: You famously conducted an experiment with toddlers that were too young to talk but who were old enough to watch a puppet show. The puppets were playing with a ball, and most of the puppets played fairly — but one did not. The toddlers not only withheld treats from the "bad" puppet, but one of them gave him a whack on the head. This study was astonishing to those of us who think that morality is related to reasoning and the ability to give reasons for what we do. Obviously, these children were too young to articulate reasons. What do you think is going on with the moral development of children?

A: These studies suggest that even babies can tell the difference between good acts and bad acts and respond accordingly. More generally, the results from our labs and others suggest that a lot of morality is bred in the bone — we are born with concern and compassion, the capacity to judge the actions of others, and even a rudimentary understanding of justice and fairness.

I should add, though, that this isn't the whole story — there is a nastier side to our natures as well. There's a lot of evidence that even the youngest babies carve the world into Us versus Them — and they are strongly biased to favor the Us. We are very tribal beings. Our natures are not just kind; they are also cruel and selfish. We favor those who look like us and are naturally cold-blooded towards strangers. And so learning and socialization are critically important if we want to transform babies into fully moral beings.

Q: And the second part of that question is: What do you think is going on with our confusion about what moral reasoning is and how we acquire it? What is it about moral development that we had so badly misunderstood?

A: I think people often get two things wrong about morality.

First, many people believe that we are born selfish and amoral — that we start off as natural-born psychopaths. As you can see, there is plenty of evidence that this is mistaken.

Second, many people believe that we are, as David Hume put it, slaves of the passions. Intelligence and wisdom are largely impotent; we're just driven by our emotions. Now, this a very tempting view of human nature given the nature of this election season. And if it's true, we should buck up and learn to live with it.

But I argue that it's not true; these dismissive claims are refuted by everyday experience, by history, and by the science of developmental psychology. We are moral animals, but we are also rational animals, and we are powerfully influenced by our capacity for reason.

Q: Psychologists and philosophers have a long history of holding that the cognitive abilities of the human baby are limited. You've written that one of the great discoveries of modern psychology is that the traditional view is mistaken. Is that mistake still slowing us down, not only in terms of our science, but in our roles as parents and teachers, nurturers of our young?

A: Not really. You can be a wonderful parent or a wonderful teacher even if you don't know the modern discoveries about the minds of babies. On the other hand, learning about the discoveries does make you look at these babies in a very different way, and I think adds a bit of pleasure to parenting and teaching.

Q: How do you study the mental life of babies? It seems almost impossible, to find out what's in someone's mind without the use of language. How do you do it?

A: Some studies study what babies look at and how long they look. Like adults even the youngest babies look longer at what interests them and what surprises them, and so you can use this as a way to explore how they make sense of the world and what they expect.

In most of our own studies, we test somewhat older babies, and although they can't yet speak, they can reach. And so we test them with puppet shows. We show babies characters who interact in certain ways — such as one individual helping another or one individual hitting another — and then see who the babies want to interact with, who they want to reward, and who they want to punish.

Q: Karen Wynn, who runs the Infant Cognition Center at Yale University, did a study that suggested that babies have the ability to do rudimentary math.

Can you tell us about that, please?

A: So here's the study: Imagine facing an empty stage. Now a hand places a single Mickey Mouse doll on the stage. A screen is placed in front of the doll to hide it from view. Then the hand brings out another Mickey Mouse doll and places it out of sight behind the screen. Then the screen is removed. As an adult, you know that one Mickey plus another Mickey equals two Mickeys, so there should be two, not one or three. Karen Wynn finds that five-month-olds have the same expectation. And when babies are shown two dolls and then one is removed, they know there should be only one left. And so babies can add and subtract.

Q: You'll be speaking in Galveston. What would you like

those who work with young children — parents, teachers, daycare workers — to know?

A: More than most, they'll be interested in the science here. Anyone who spends a lot of time with babies and children will be fascinated by the questions of how their minds work.

Q: OK, I've got to ask. Why did you call your book, "Just Babies"?

A: My wife thought up the title, and it means to be playful, with a double meaning. "Just Babies" can express a reasonable skepticism about the abilities of these tiny creatures — what do you expect of them, they're just babies! But of course "just" also derives from justice — as in "a just society" — and so the title captures one of the

main arguments of the book, which is that we are born as moral creatures. We start off
as just babies.

(Published as "Q&A: Paul Bloom set to speak Nov. 2, The Galveston County Daily News, *Oct. 25, 2016)*

3. Susan Landry — March 7, 2017

Brain development in childhood
By William J. Winslade

Several years ago a 10-year-old boy was given a gun. When his father, an emergency room physician at UTMB, came to pick him up from his mother, who was divorced from the father, the boy climbed into the back seat. He fatally shot his father through the back seat. This is only one of the many tragic episodes that result in a child's being committed to the Giddings State School in rural Texas.

Many juveniles who have committed lesser crimes and some who have a long history of serious juvenile delinquency end up as adults in prison. But not before they damage the lives of others, as well as their own.

Although many factors may contribute to both good and bad influences on the cognitive, emotional and social development of the young child, we all know that parents and teachers play a key role in positive and sometimes negative habits and actions of children from early infancy to adulthood.

One might think that raising a young child just requires love and common sense, rules and strict discipline — and everything will be all right and work out for the best. But it is not so simple or easy.

What then is necessary, even if not sufficient, to provide the best environment for the behavior and brain development for the young child?

Susan Landry and her colleagues at the Children's Learning Institute at the University of Texas Health Science Center at Houston have extensively researched the factors that parents and teachers can provide for the optimum development — cognitive, emotional, personal and social capacities and needs.

It is what she had described as responsiveness to children's needs by parents and teachers. Each child is a unique, complex and evolving person.

Responsiveness refers to cognitive, as well as the linguistic, imaginative, emotional, social and interpersonal opportunities for development and learning.

Her research demonstrates that responsiveness requires child-centered care that is focused on the specific needs, interests, talents and personalities of each child viewed as a unique individual. That type of responsiveness is not easy to discern, especially if the parents or teachers think they know or even always know what is best for a particular child. The only way to determine that in a particular case and for a particular child is to avoid one size fits all thinking.

We now know, or should know, that different children have different thoughts, fears, likes and dislikes, and each child needs to be treated accordingly.

From my experience as a psychoanalyst I am convinced that every person I have seen for therapy has a wholly unique inner and outer life story. I think that it is the same for the young child even though his or her life story is shorter — even though it is likely to be much more complex than one might think. That is why responsive parenting or teaching should take the individuality of each child's needs and capacities into consideration.

As a teacher of undergraduates, graduate students and professional school students, I see each student with different interests, capacities, skills, insights and strong and weak areas. And the same is true for each of us, not just young children.

So we must try to remember and to respond to young children with an appreciation for their evolving unique identities and personalities.

(Published as "Talk slated on childhood brain development," The Galveston County Daily News, *March 3, 2017)*

+++

Q&A with Susan H. Landry

Q. You have written about the importance of "responsive

parenting." Please tell us a bit about that. Why is it so important?

A: Responsive parenting includes caregivers responding to children's signals (e.g., needs, interests, requests) promptly, sensitively and in ways that are directly linked to what the child is signaling. It includes the provision of high levels of warmth and the acceptance of children as unique individuals.

Responsive parenting is most effective in supporting young children's development if it integrates warmth and sensitivity with rich cognitive stimulation, including conversations with young children that include rich vocabulary and concepts about children's world (e.g., how things work, why objects go with certain actions).

Responsive parenting is thought to be important for children's long-term development because when implemented consistently children experience support for their individual needs and thus feel valued. This, in turn, promotes a willingness for children to engage in positive interactions with their caregivers. Recently, a longitudinal study found that responsive parenting in the first three years of life was directly associated with positive life outcomes for adults at 32 years of age.

Q. Is it possible to measure the effects of responsive parenting? And how is that done? Do you measure a child's behavior or the structures of the brain?

A: To date most studies measure

the effects of responsive parenting on children's development by measuring the child's behavior. Responsive parenting explains children showing better outcomes in language, self-regulation, and social and emotional skills.

However, recently studies have measured the effects of responsive parenting by examining structures of the child's brain. Young children whose caregivers show higher levels of warmth and responsive support show differences in brain volume in structures such as the hippocampus and amygdala, areas of the brain important for the management of stress.

As this relation between responsive parenting and brain development can be detected in the first three years of life, the implications for children's long-term outcomes is critically important.

Q. Is it possible to teach the skills, techniques and strategies of positive parenting? How could we best go about that?

A: It is possible to support caregivers to learn the strategies associated with responsive parenting. Programs that have been successful in doing this include some key characteristics such as involving caregivers actively in the process of learning to use different behaviors with techniques such as using video examples of believable parents using responsive strategies and then talking about why the strategies were helpful to children.

One effective responsive

parenting program shows video examples and then has caregivers video themselves using a responsive strategy with their young child. Together the caregiver and their parenting coach critique the video, with the caregiver pointing out behaviors they thought were helpful for their child.

This program can also be implemented remotely with caregivers viewing the sessions on tablets and then uploading a video that they can watch and discuss with their remote coach. Remote programs such as this one are new but preliminary results are encouraging.

Q. We all know that the world is not full of perfect parents. What can educators do to fill in the gaps when parents don't help their children to develop?

A. There are many ways educators can responsively support very young children's learning and development. The same warm and reciprocal interactions that are important in the home need to occur between young children and teachers and staff in center-based early childhood programs.

Integrating responsive support with rich cognitive stimulation is just as important in the classroom as in the home environment. As the early childhood period is one of the most critically important periods of a child's life for optimal brain development and learning, it is important that adults provide responsive support in every environment where young children spend time.

Ideally, communities will work to align responsive quality support for young children across home and center-based environments.

Q. What do you hope that people who attend your lecture in Galveston will come away with? Is there a key point you want to make?

A. A key point of this lecture is how critical the first years of life are for children's brain development and learning and an understanding of what children need in these early years to better assure they will be successful in life.

If we wait until pre-K to address developmental problems that are the result of low levels of responsive parenting, including poor cognitive stimulation, then it is often impossible to totally ameliorate learning delays and deficits that can have devastating effects for children's outcomes. There are solutions but they require funding, persistence and a commitment from many.

(Published as "Q&A with Dr. Susan H. Landry," The Galveston County Daily News, *Feb. 28, 2017)*

4. Silvia Bunge — Sept. 13, 2017

Changes in adolescents
By William J. Winslade

Silvia Bunge, an acclaimed neuroscientist, on Sept. 13 will deliver the fourth talk in the Robert and Russell Moody Lecture Series: "Children and Adolescent Brains: How Do They Change?"

I started thinking about my own childhood and adolescence.

When I was 22 months old, I fell from a second story porch landing on my head on a concrete pavement. A fractured skull and a traumatic brain injury kept me in the hospital for five weeks.

In 1943 the only thing the doctor did was drain the fluid and remove the bone fragments from my skull.

I was incredibly lucky to suffer no permanent disabilities.

A couple of years later my parents moved from Chicago to a small town in central Illinois, where I spent my childhood and adolescence. Although I remember all my teachers and friends, I remember little about learning anything, even though my grades were much better than my conduct. I do recall that during those years I spent as much time as possible outdoors. I read only what was necessary for school.

In high school I was busy with sports (basketball, track and golf). I also played saxophone in the school and city bands. In addition I worked first mowing lawns and later making pizzas at the small pizza place operated by my mother and one of her friends.

Schoolwork was something I did because I had to do, but none of it was very interesting. I did little reading, and most of my thinking was about what I was doing or going to do the next day.

In retrospect, it is obvious that my college decision-making also exemplified a general lack of foresight and planning. My action-oriented adolescence had little room for school, reading, reflection. It had very little intellectual interests or self-awareness.

For me a great intellectual awakening arose when I stopped falling asleep in my introduction to philosophy class. I had never even heard of philosophy until I took this class. But the lectures and the book written by the teacher fascinated me.

From that point on I was obsessed with philosophy and became much more interested in reading novels and pondering the vast array of philosophical issues my teacher introduced, such as aesthetics, ethics, philosophy of mathematics and science, and philosophy of religion. Either I had never even thought about such things or had taken for granted what I had heard in church.

At that point my intellectual interests and curiosity led me to abandon my tentative plan to attend law school and to go to graduate school in philosophy. All of my subsequent intellectual and professional interest and education in law and psychoanalysis grew out of my college experience in my philosophy classes.

My childhood and adolescence were filled with sports, music, work and had very little to do with academics or learning. But the study of philosophy awakened and stimulated a part of my brain that had not previously been activated and cultivated.

As I look back on my childhood and adolescence, I am somewhat amazed about how little I thought about anything other than sports, work and socializing.

I still wonder how my brain changed. I hope Silvia Bunge helps us all learn more about the brains of children and adolescents.

(Published as "Acclaimed neuroscientist set to speak Sept. 13, The Galveston County Daily News, Aug 9, 2017)

+++

Q&A with Silvia Bunge

Q. You are known for your work on the neural development of children and adolescents. How did you get interested in that topic?

A. I was interested in studying how the brain works and joined the lab of John Gabrieli as a graduate student in 1997. This was the early days of functional MRI, and nearly all of the studies had been done in adults. I didn't have a particular interest in development, but I understood that studying how the brain matures would be a good way to glean insights into how it functions in adulthood. And then I gradually discovered how fascinating brain development is in its own right — and how important it is to understand it.

Q. In a nutshell, how does the brain change as the ability to reason emerges between the ages of 6 and 19?

A. The basic brain architecture is already in place by age 6, meaning that if you look at a 6-year-old's brain with the naked eye, it wouldn't look all that different from an adult's brain. But if you take careful measurements with MRI, you will find that the connections between the brain regions that are involved in reasoning get stronger over childhood and adolescence.

Q. You've suggested that the reasoning abilities of kids at those ages is a good predictor of what they'll be able to achieve in college and in demanding careers. Who might benefit from this talk? Is it mainly for teachers, students, parents?

A. Every one of us has to reason our way through problems in our lives, as well as in school and at our jobs. However, people vary widely in their reasoning skills, and so it's important to understand whether, when, and how we can improve at it. This line of research is relevant for people who are curious as to how our brains change, and how these changes support cognitive development.

Q. It's almost as if exercising the mind can develop the brain and make you a better student, just as lifting weights and running sprints can develop the body and make you a better athlete. Athletic training certainly has changed with scientific investigation. Do you think we'll see similar changes in cognitive training, based on new findings in neuroscience?

A. Some researchers don't like the analogy between exercising the mind and exercising the body, but I think it's quite apt. In both cases, prolonged and repeated training is necessary. Benefits are hard-earned, and fade after you stop "working out." And the benefits are quite specific, just like with weight-training: you can't lift weights with your arms and expect your calf muscles to get stronger. Likewise, you can't practice one kind of cognitive skill and expect to see benefits across the board. Another reason I like the analogy is that — as you note — athletic training has improved as understanding of the body has accumulated. We're still at the early stages of figuring out how best to boost learning and brain plasticity. Researchers are experimenting with various approaches, for example combining cognitive training with aerobic exercise training, or with transcranial magnetic stimulation, or even with pharmacology, and there have been some promising early results.

Having said all of this, I am not generally an advocate for the computerized, "kill and drill" approach to learning that is at the heart of cognitive training. I do however think that it could be a helpful adjunct to traditional education for children who are struggling in school and need to strengthen foundational cognitive skills.

Q. Incidentally, some of us are impressed that researchers in the field of neuroscience give public lectures, trying to inform the public about an important but difficult topic. Is helping the public understand this field a matter important to you?

A. It's vital that scientists engage with the public, and I try to do my part. Some do it way more than others — and that's OK. We need people in the "trenches" who are doing good research and sharing it with the scientific community, and we also need people on the "front lines: who can communicate effectively and accurately with a broader audience.

Q. Your work reflects an interest in how mental habits affect brain function. Can you give us an example of a mental habit that is worth developing during adolescence?

A. I worry that teenagers don't read books for pleasure very much anymore. Reading requires sustained attention (sitting still and focusing on the printed word when there are so many distractions), working memory (keeping track of who's talking or what's happening from one page to the next), and long-term memory (remembering the plot line when you pick up the book again after some time away), and it provides the opportunity to build vocabulary. Also, former

President Obama said that he liked to read novels because they help him to develop empathy.

Q. What would you like people who attend the lecture to come away with?

A. Reasoning is essential for learning, and learning is essential for reasoning. We are beginning to understand how the brain reasons, why we get better at it over childhood, and how our minds and brains change when we practice our reasoning skills.

(Published as "Neuroscientist Silvia Bunge set to speak Wednesday," The Galveston County Daily News, Sep 8, 2017)

5. Robert Duke & Arthur Markman, Oct. 25, 2017

Is there a trick to happiness?
By William J. Winslade

What is happiness? What makes a person happy or unhappy? To what extent is our happiness or unhappiness something a person can control or change? To what extent is our disposition to be happy or unhappy a result of our genetic inheritance? These are just a few of the questions that Robert Duke and Arthur Markman, professors from the University of Texas at Austin, will discuss Wednesday in their Robert and Russell Moody Lecture at Moody Gardens.

In this commentary I will offer a

few of my own thoughts about happiness. In Plato's book *The Republic,* Socrates suggests that happiness arises from the harmony of a person's soul. One might say that people are happy when they are in tune with themselves and unhappy when they are out of tune.

But what causes happiness or unhappiness? I am happy when I feel energetic and can think clearly. But I am unhappy if I am sick or tired or frustrated. Success produces happiness; failure engenders unhappiness. I am happy when friends and family are doing well but unhappy if they are not. Serious illness or death of a loved one or close friend can make us unhappy. We mourn their loss or we may become depressed or even melancholic. But if they recover from illness we are happy for them.

We are happy when we feel healthy, strong and successful, but unhappy when sick, weak and unproductive. Sometimes we can change our bad mood. Sometimes we can regain a happy disposition through problem solving, exercise or therapy. Pain may make us unhappy, but therapy — physical or mental — can transform our mood and outlook.

Some people seem to have a greater capacity for happiness, while others may typically see the dark side of things. Others fluctuate more in their moods based on their mental state or external events. Positive thinking promotes happiness whereas negative thinking pulls us toward unhappiness.

What is clear is that we do have some degree of control over mental and emotional states. Some people are more or less able to stay in tune with themselves than others.

Happiness and unhappiness then partly depend on how we feel about the world around us and partly depend on whether good or bad things are happening. We may not be able to control external events — fires, floods, wars, etc. — but we have some control of our responses. Likewise, we feel happy when things go well and are happy for others. We should be aware that we may not be able to control even the good or bad things that happen, but we always have the possibility of controlling ourselves.

Some people are more resilient

than others. Others are prone to having fluctuating feelings that may be proportionate or disproportionate to what happens to us or to others. We have more capacity to control how we respond than to control external events. It may take extra effort to remain happy or to overcome unhappiness, but the good news is that our attitudes are often under our control. We can try to stay in tune and retain the harmony of our souls.

(Published as "Is there a trick to happiness?" The Galveston County Daily News, *Oct. 23, 2017)*

+++

Q&A with Art Markman

Q. You guys are known for "Two Guys on Your Head," a radio show that examines some of the baffling aspects of human behaviors. First, tell us about the show. What "Click and Clack" does for cars, your show does for minds — is that an accurate description? What are you trying to do with public radio listeners?

A. That is right. We want to bring a little bit of the science of psychology into people's week. The beauty of psychology is that most people can find an immediate application of these principles to their lives. We aim to make the science relevant and fun. The show is done in a light-hearted tone. The science may matter a lot, but it is still fun.

Q. You both have doctorates in different areas: one with expertise in psychology and marketing and one with expertise in music and learning. Is that part of the success of the show — the start from different viewpoints?

A. I think that the success of the show comes from a combined 50-plus years of experience in our fields. Both of us are generalists in our approach to psychology. Despite the particular areas of research we have focused on individually, we both read widely within the field and think integratively about how the mind works. That enables us to tackle a wide variety of topics.

Q. The Moody Lecture Series in Galveston has focused on the

development of the brain, and we've had speakers talk about infants, toddlers, children and younger adolescents. How about the older kids — high school students and college freshmen? How do they make decisions?

A. The fascinating thing about older kids is that for many of the important decisions in their lives, the process doesn't feel like a stereotypical decision. When someone goes to the store to buy tomato sauce or perhaps a video game, they are confronted with a set of options. They compare and contrast those options and select items based on the most important differences. With choices like where to go to college, what to major in, or what field to work in, students often go through options sequentially. They try a particular school / major / career choice on for size for a while and stick with it if they like it. Otherwise, they switch to another possibility — perhaps influenced by a class, mentor, or other experience. It is rare that students report that they were actually choosing from among a set of options.

Q. And your talk is going to be about happiness. Is it something we create?

A. Happiness is a mix. About 50 percent of the difference between people in levels of happiness results from genetic factors. So, people don't have complete control over how happy they are overall. That said, people can also engage in many habits that increase their happiness. For example, focusing on positive memories and expressing gratitude are habits that can make people happier. People can also manage their environments to make them happier. Minimizing commute time to work and hiring people to do things people don't want to do will make them happier.

Q. Also, you have a book out called *Brain Briefs*. Please tell us about that.

A. *Brain Briefs* is like the show. It is a collection of short chapters organized around topics that range from deeply important (like can we make ourselves happy) to not-so-important (like why people like kitten videos so much). The aim was to teach a little psychology in a way that is fun and can be consumed in short doses.

Q. Is the talk in Galveston going to be like the show?

A. Yes. Whenever Bob and I get together, we just start talking. In many ways, these talks follow the format of the show. We also like to leave a lot of time for people to ask questions. After all, people don't always have the chance to ask a psychologist or two any question they want.

(Published as "Q&A: 'Click and Clack' of the mind to speak next week,' The Galveston County Daily News, *Oct 15, 2017)*

6. Walter Glannon, Feb. 7, 2018

How do you enhance your mind?

By William J. Winslade

Do you want to improve your mind? To think more clearly? To have a better memory? To stay focused rather than being distracted?

There are many things that you can do such as improve your diet, exercise, take advanced education classes, take yoga or relaxation exercises, or read challenging literature. Another possibility is to take cognitive enhancement drugs.

On Feb. 7 at Moody Gardens, philosopher of mind and neuroscience Dr. Walter Glannon will discuss a wide range of drugs that provide cognitive enhancement.

In this commentary I discuss how psychotherapy and psychoanalysis can also help a person gain a specific type of cognitive enhancement: self-awareness about your feelings and emotions that you may not realize are influencing your behavior.

As a licensed psychoanalyst I do not offer advice. Instead, I simply ask a person who consults me to talk about whatever comes to his or her mind.

After listening to a person over time, to the thoughts, feelings, dreams, recollections and events from the past or present, I may ask a question or make a comment that will help a person discover an insight, a hidden feeling or an emotional conflict.

But it is the person in psychoanalysis rather than the analyst who may become self-aware of something significant which the person had not realized before.

This type of self-discovery and self-knowledge is a special type of cognitive enhancement that may or may not influence the person to think, feel or act differently. But this type self-knowledge may then help a person to better understand past, present and future episodes in the person's life.

Psychoanalysis was once described as being the exploration of a cave in which the person in analysis is the explorer and the psychoanalyst follows behind the explorer and occasionally shines a flashlight on a dark corner to help the person find his or her way about.

What the person does, if anything, with the new self-awareness or self-knowledge is left up to that person.

Although psychoanalysis is not usually described as a way to achieve cognitive enhancement, it is my belief that it is one of the many ways that persons can improve their minds to help them better understand themselves.

(Published as "How do you enhance the mind?" The Galveston County Daily News, *Jan. 25, 2018)*

+++

Q&A with Walter Glannon

Q. We seem to be on the verge of

being able to enhance our cognitive abilities with the use of drugs and therapy. Do you think that's wise?

A. Enhancing cognitive abilities can benefit us in many ways. By improving working memory and other cognitive functions, it can enable us to perform more cognitive tasks more efficiently and effectively. Whether it's wise or not depends mainly on two things, one medical, the other social. First, the risks of long-term use of Ritalin or the wakefulness drug Modafinil on the body and brain are not known. Studies of short-term effects have shown that there are cognitive trade-offs with these drugs. They can improve some functions but impair others. Second, enhanced cognition could increase competition in schools, universities and the workplace and exacerbate what is already a high level of harmful psychosocial stress.

Q. What about a healthy college student who has no cognitive disorders who is tempted to take a drug — let's just say Ritalin — to give him a competitive advantage. Is that fair? We have a complicated system to prevent athletes from using drugs that enhance performance.

A. It does seem unfair for a student with normal cognitive functions to take a cognition-enhancing drug to gain a competitive advantage over other students who do not take them. The difficult question is whether we can distinguish these drugs from other drugs, such as caffeine.

Students find many ways of enhancing their cognitive capacities. Are cognition-enhancing drugs significantly different from other means of enhancing ourselves? Also, taking Ritalin or another drug alone will not guarantee a positive outcome. Other factors such as motivation and being conscientious are important as well.

Q. What about drugs for Alzheimer's? Few of us would have any qualms about taking a drug to alleviate symptoms of that terrible disease. Do you see any ethical concerns?

A. Some people in the enhancement debate argue that there is no clear distinction between therapy and enhancement. It's just a matter of degree relative to a "normal" level of functioning. But others argue that there is and should be a distinction between therapy and enhancement. Drugs for Alzheimer's that improve memory are therapy, rather than enhancement, because they are for a disease in which memory falls below a normal level. Improving memory function to approximate a normal level is different from improving memory that is above this level. Drugs that improved memory in some Alzheimer's patients may make them more aware of their condition, which could cause distress and thus harm them. But this is speculative. No studies to date have shown this effect.

Q. Perhaps the impetus behind the interest in performance enhancers is self-improvement, which doesn't seem like such a bad thing. Most of us invest years in getting an education. We all favor that kind of self-improvement. Do you see a clear line where our efforts at self-improvement go astray?

A. Many people argue that taking enhancers is part of self-improvement and thus not objectionable. The drugs are necessary for us to develop our natural physical and mental capacities to the fullest extent. But this raises the question of how much of what we do is really up to us, and how much of it is the result of things that are not intrinsic to our natural selves. It raises questions about autonomy and authenticity.

(Published as "Q&A with Walter Glannon," The Galveston County Daily News*, Jan. 27, 2018)*

7. Nicholas Schiff, April 4, 2018

How scientists see consciousness
By William J. Winslade

Human consciousness is an amazing but also a mysterious phenomenon. When we are awake we take our consciousness for granted and when we wake up from a deep, dreamless sleep we expect to return to consciousness.

But our consciousness depends upon our brain function, and our brains are vulnerable to injury, disease or other insults, such as a chemical imbalance in our bodies caused by drugs, toxic fumes or an allergic reaction. When our brains become seriously dysfunctional, we may lapse into a coma. A coma is a loss of consciousness that may last for hours, days or even a week.

If a person in a coma does not recover consciousness, he or she may die or lapse into what is called a vegetative state, that is, a condition of unresponsive unconsciousness. A person in a vegetative state — with adequate life-support technology, such as a respirator and medically administered nutrition and hydration — may survive for a long time. For example, Rita Green, a young nurse who had a reaction to an anesthetic, remained in a vegetative state for over 40 years.

You may remember Terri Schiavo, who was in a vegetative state for 15 years until her husband was authorized by a Florida court to discontinue her nutrition and hydration despite an acrimonious national controversy.

Whenever a person is in a long-term vegetative state, it often creates difficult decisions for physicians, hospitals and family or other designated decision makers. If a person in a vegetative state has not told anyone what he or she would want, it typically causes indecision and conflicting opinions about the use of long-term life support.

There's an opportunity to learn more about the loss and recovery of consciousness from a vegetative state and surprising discoveries about some persons in a vegetative states.

Dr. Nicolas Schiff, a leading expert on the loss and recovery of consciousness, will give a free public lecture at Moody Gardens.

One of Dr. Schiff's surprising discoveries is that some persons in a vegetative state may recover consciousness through deep brain stimulation — the stimulation of the brain by electricity.

Another startling discovery made by Dr. Schiff is that some persons who appear to be in a vegetative state and who cannot respond verbally or physically to commands may be able to respond with their brains. This is what Dr. Schiff has called "covert consciousness."

To learn more about the

vegetative state and these extraordinary discoveries made by Dr. Schiff, come to the lecture.

(Published as "Talk on the mysteries of human consciousness set for April 4," The Galveston County Daily News, *March 28, 2018.)*

+++

Q&A with Nicholas D. Schiff

Q. You've suggested that some people who have been diagnosed as being in a vegetative state are actually in a minimally conscious state. What's the difference, and why is that distinction so important?

A. The distinction is that patients in the minimally conscious state

(MCS) show definite evidence of awareness and have a very different natural history pattern of recovery. Many MCS patients are now known to continue to recover for many months and even years with a significant fraction of patients remaining in MCS after a severe brain trauma for a few months, reaching independent recovery and re-employment after long recoveries.

Q. In a famous case, a Texas man named George Melendez was in a minimally conscious state after almost drowning. He lost the ability to speak. But because he was groaning at night, his mother gave him Ambien, a common sleeping pill. He started talking and, with time, was able to feed himself. But he still needs the drug to be able to feed himself. What does that tell us?

A. It is a very clear example of how much brain function the severely injured brain may continue to harbor without evidence at the bedside of this latent capacity. It also shows us that after severe injuries it is often necessary to reorganize the dynamic activity of the brain to regain more optimal function, much like cardiac pacemaking restoring the "pump" function of the heart.

Q. You were involved in treating a man who was severely beaten in a mugging. Most of the time, the man was unresponsive, but brain scans revealed that he did recognize his mother's voice. You suggested that he be treated with deep brain stimulation — a technique used on people who suffer from Parkinson's disease that involves implanting electrodes in the brain. The treatment was successful, and the patient's mother spoke of his coming back — he regained his ability to eat, to stay awake during a conversation, to express pain and emotions, including love. What does that case tell us about the rules of healing?

A. This case tells us that we can intervene late in the course of what appears to be a stable and permanent level of recovery and make a significant functional difference. We found that deep brain stimulation produced not only these effects, but it initiated an ongoing process of recovery despite a six-year period of lack of change without the intervention. The findings also confirm a strong prediction about the critical role of the brain network we stimulated in the recovery process.

Q. If we understand the biological process of how some people recover from injuries, does that imply that we might be able to understand how people could recover from disease? Do you see a day when medicine might be able to bring people back from Alzheimer's disease?

A. Understanding the rules of the game underlying recovery from severe structural brain injuries will have widely generalizable applications. Our findings in studies with many different types of structural brain injury support this view. While the causes and nature of a process like Alzheimer's disease take their origin in the widespread loss of neurons (brain cells), if we get to interventions to stop or slow the disease the principles we are uncovering in the structural injured brain will have important implications. In another degenerative disease, Lewy Body Disorder, that affects almost as many Americans as Alzheimer's, the application of the principles we are uncovering have immediate application, and I have seen several patients with this type of degenerative illness do better with an approach derived from our evolving understanding of structural brain injury and mechanisms controlling states of consciousness.

(Published as "How do 'vegetative' people come back?" The Galveston County Daily News, *April 3, 2018)*

8. Daniel Wikler, July 12, 2018

How do you define death?
By William J. Winslade

How do you know if someone is dead?

It was easier when the only way to be sure was when the person did not breathe or have a heartbeat. But now "brain death" is a widely accepted criterion for a physician to use in declaring that a person is dead. But what is brain death and how is it determined?

Despite the fact that brain death is accepted in nearly all American states, confusion and controversy still occur.

For example, a patient on a respirator appears to be alive because he is breathing and has a heartbeat, but the doctor tells the family that the patient is brain dead. Family members are worried that the declaration of death is premature. The patient has signed an organ donor card. Family members wonder whether the doctor is interested in taking the patient's vital organs for transplantation.

They don't really understand the idea of brain death. They don't understand because they can't see brain death and they are not sure whether the doctor is being truthful.

Brain death can be a controversial topic. If you want to learn more, come hear a world-renowned expert, Daniel Wikler.

As "staff philosopher" for the President's Commission on Ethical Problems in Medicine under Presidents Carter and Reagan, he was an author of its report "Defining Death," which resulted in the redefinition of death in nearly all states.

Wikler is the Mary B. Saltonstall Professor of Ethics and Population Health at the Harvard School of Public Health. He has contributed to bioethics both as an academic and in public service.

He served as the first "staff ethicist" for the World Health Organization and remained a consultant to several programs, most recently its consultative group on ethical dimensions of universal health coverage.

(Published as "World-renowned expert to discuss brain death," The Galveston County Daily News, *July 10, 2018)*

9. John Paul "Paulie" Gaido, Sept. 26, 2018

How should a good life end?
By William J. Winslade

Ever since 1976 when Karen Quinlan was put on a respirator for months while she was irreversibly unconscious, Americans have wondered about the proper use of life-sustaining technologies such as respirators and medically administered nutrition and hydration.

Many, if not most, adults in Texas have heard of the Texas Natural Death Act or the medical power of attorney for health care. But do you really know what your legal rights are?

Have you actually discussed what these laws do and do not permit you to do? Even though the first Natural Death Act was passed in Texas in 1977, it has been modified numerous times through the legislative cycles.

Even adults who already have an advance directive may need clarification or an update.

An advance directive will document what you would want if you have a terminal illness or an irreversible condition in which a decision must be made about the use of life-support technology. If you are incompetent or unconscious, an advance directive can protect your preferences.

You have the right to stipulate your preferences and/or designate the person who can speak for you.

But if you really want to understand your rights and why it is a good idea to exercise them, you need to come to The Robert and Russell Moody Lecture on Sept. 26 to hear Paulie Gaido, who is not only a lawyer and a medical humanist but also an expert on advance directives. He can tell you why and how to protect your rights and make your desires known if you are unable to speak for yourself.

He will tell you not only what you can do, but, more importantly, why you should do it.

+++

Q&A with Paulie Gaido

Q. Most of us, at a certain age, are familiar with the idea of an advance directive. Why are they so important?

A. Your serious illness is traumatic enough for your family without forcing them to make medical treatment and care decisions on your behalf, should you be unable to do so. Taking the time to express, in writing, your medical treatment and care preferences is, in my opinion, a good way to say "I love you and don't want to burden you with the responsibility of making decisions on my behalf without adequate guidance from me." Telling your spouse: "You'll know what to do" or "I am comfortable with whatever decisions you make" significantly underestimates the difficulty in making decisions, especially when the cost of medical treatment isn't completely covered by insurance, or has significant side effects, yet is capable of sustaining your life.

Q. And why do we have such difficulty in writing them?

A. First, few of us like talking about injury, illness or disease so serious that we would be incapable of making our own medical treatment and care decisions. Second, it is admittedly very difficult to accurately foresee the future. Nevertheless, you can make medical treatment and care decisions in advance and delegate the authority to a family member or trusted friend to knowledgeably modify those decisions should the future be significantly different than anticipated.

Q. In your dissertation, you recommended starting with a statement of your values — the things that define you as a person. Why is that such an important strategy?

A. Advance directives typically communicate the what of your medical treatment and care decisions. In order to maximize the moral authority of those decisions you need to communicate the why, especially that those decisions are anchored in, and totally consistent with, your values

Q. What point do you plan to make in your talk that you hope most people will take to heart?

A. Your medical treatment and care decisions, even when made in advance, have undisputed legal authority. Without corresponding moral authority, however, those decisions could, under certain circumstances, be ignored or simply overridden by those who are convinced that they know better than you do what is best for you. Maximizing the moral authority of your decisions is not difficult. The first step is writing a personal statement of values and a personal statement of healthcare goals. Both are, in my opinion, surprisingly easy to write, and I am giving everyone who expresses an interest a copy of my own as a sample and starting place.

(Published as "Gaido to open fall's Robert and Russell Moody Lecture Series," The Galveston County Daily News, *Sept. 18, 2018)*

10. Joseph Fins, Oct. 24, 2018

Different kinds of brain injuries
By William J. Winslade

Traumatic Brain Injury is a major cause of injury, death and health care costs in the United States.

Traumatic Brain Injury is caused by falls, especially among the elderly and young children, vehicle crashes, assaults, and sports injuries. The health care costs and economic loss are enormous.

Emergency room visits, hospitalization, and rehabilitation programs are required for many thousands of patients every year, yet our prevention efforts fall very short of the need for more effective ways to decrease Traumatic Brain Injury.

Even when victims of Traumatic Brain Injury survive their initial injuries and receive emergency medical care, they do not receive the adequate follow-up care that they need. Unfortunately, there is a large population of survivors whose needs and rights are not properly protected.

Dr. Joseph Fins, a physician and medical ethicist from Cornell University Medical School, has interviewed a thousand families who have a family member with a Traumatic Brain Injury who has not received adequate care. All too often such patients do not receive adequate rehabilitation and are instead transferred to nursing homes that are ill equipped to meet those patients' needs.

Dr. Fins has documented the plight of such patients in his recent book, *Rights Come to Mind*. He presents a powerful and moving account of why Traumatic Brain Injury patients not only do not get the treatment that they need but are also deprived of their civil rights.

This lecture will not only inform listeners but it will also explain why failure to meet the needs of patients with Traumatic Brain

Injury is a violation of their civil rights.

This is a topic that every adult, parent and child needs to not only hear about but also help to promote prevention and to restore the rights and meet the needs of Traumatic Brain Injury patients.

(Published as "Moody lecture series continues Wednesday," The Galveston County Daily News, *Oct 21, 2018)*

+++

Q&A with Joseph J. Fins

Q. Up to 40 percent of the cases involving minimally conscious states and vegetative states are misdiagnosed. For us laymen, what is the difference? And why is this so important?

A. People who are vegetative are not conscious. They don't feel pain. They are unaware of their environment. But their eyes are open. They have sleep-wake cycles. A minimally conscious patient may look most of the time like a vegetative patient but they are conscious. They are aware of the environment and of others and can feel pain.

Q. You have described consciousness as a civil right. How will thinking about consciousness in that way help people who have suffered injuries?

A. People in minimally conscious states, if they don't progress quickly, end up in nursing homes, where they typically don't get the standard of care they deserve. They are ignored and they are segregated — and I use that word in the context of the Civil Rights Movement. They are getting separate and unequal care. Their rights are being violated. The Americans with Disabilities Act says that people should not be segregated and should be maximally integrated into society. We understand that means we improve the sidewalk to help a person in a wheelchair get to work. But we don't often think of people who are conscious but minimally so and the assistive devices that could integrate them back into society.

Think of the indignity of being in a nursing home bed, being ignored, being turned without pain medication because everybody thinks you're not there. ... I think it's the next wave of the Civil Rights Movement.

Q. You've argued that it's vital to support families of people who are in a minimally conscious state. What kinds of things should we be doing?

A. First, we have to get the diagnosis right, and we have to get the diagnosis right over time. Brain states can change. We have to make sure they are credibly diagnosed so that families are making informed decisions about what their choices are.

I'd also like to improve access to rehabilitation. We're not providing adequate care. We know more. We can do better. But that better is not accessible. That puts a tremendous strain on families.

I think families also need to have respite care. They need to have their own souls restored. There may be a role here for hospices for psycho-social support to these families who are in a state of almost constant anticipatory grief or bereavement.

I also think families need a tremendous amount of education. Details make a big difference. We really have to educate the population. That's why I am grateful to the Moody Foundation.

Q. Where does your passion for this research come from?

A. It bothers me that people who are conscious are treated so badly — as if they are not there. The brain is resilient and people do recover. They deserve a chance. It seems to me wrong that we spend $600,000 to $800,000 to get people out of the hospital and then we don't spend money to make sure that they can return to some semblance of a life. As a scholar and a scientist I'm dispassionate. But I think it's a moral wrong, and I want to right it.

(Published as "Q&A with Dr. Joseph J. Fins," The Galveston County Daily News, *Oct. 18, 2018)*

11. David Van Essen, Feb. 6, 2019

Two more ways of thinking about the brain
By William J. Winslade

Two internationally known

scientists are coming to Galveston to talk about new research on the human brain.

On Feb. 6, David C. Van Essen, a principal investigator with the Human Connectome Project, will talk about a map of the brain that made worldwide news.

On March 6, Norman Doidge, the best-selling author of *The Brain That Changes Itself,* will talk about how the brain heals from injury or disease.

Van Essen and Doidge are the final speakers in the Robert and Russell Moody Lecture Series, which has explored consciousness and the lifecycle of the brain.

The whole goal of the lecture series is to open the door to anyone interested in learning something about their brains. The lectures are free and open to the public. It's a rare opportunity for anybody living in Galveston to learn from internationally known brain researchers.

Van Essen, a Harvard-trained neuroscientist, has described himself as "a cortical cartographer." In 2016, he was a leader of a team that published a map that detailed 180 regions of the brain — 97 of which had not previously been known.

In his lecture in Galveston, Van Essen will talk about the development of the new brain map and will outline what these new findings tell us about how our brains work.

He began his work in the 1960s, using a pencil and tracing paper to try to give a map of the areas of the brain that were then known. He's followed that interest through revolutionary changes in computers, imaging technology and neuroscience.

Doidge, a psychiatrist and a psychoanalyst, emphasizes the neuroplasticity of the brain — that is, how the brain is able to change to heal and compensate for injury or illness. He will present cases that illustrate how neuroplasticity can explain the ability of the brain to heal itself.

While both of these scientists do highly technical research, they will present their talks in ordinary language that any interested child or adult can understand.

These lectures will provide information and insights for anyone who is curious about the lives of our brains.

(Published as "Lectures on human brain set for Feb. 6 and March 6," The Galveston County Daily News, *Jan. 12, 2019)*

+++

Q&A with David Van Essen

Q. The map published by the Human Connectome Project in 2016 is a milestone. It detailed 180 regions of the brain, 97 of which hadn't been known. But you started mapping the brain with pencil and tracing paper. How has your idea of an image — or a map — changed?

A. A useful analogy is to think of the maps of the Earth. Some show mountains and rivers and other geographical features. Others show countries, states and other political subdivisions. Each serves a different purpose. When we make maps of the folded cerebral cortex, these are analogous to geographical earth maps. Our maps of brain function are akin to maps of political subdivisions.

Early earth maps were quaint and amusingly inaccurate, but now we enjoy the extraordinary precision of Google Maps. Brain maps have dramatically improved as well, though not to the same degree. In both domains, map-makers share similar fundamental objectives, but the tools continue to evolve.

Q. You found your field of study when you came across a book on the brain after your sophomore year at Caltech. What advice would you give to students who are interested in science but haven't found a field that really grips them?

A. I was interested in science from early childhood. In high school, I was fascinated by math and physics. In college, I realized my brain wasn't really "wired" to become a theoretical physicist. I became a chemistry major instead, but chemistry didn't ignite the kind of passion that sustains a career. By good fortune, I chanced across a book about the brain and immediately became hooked.

One lesson is that success in science requires intense work but also a fierce passion for your area of study.

Q. Some scientists have expressed confidence that we'll come to understand consciousness just as we have come to understand the human genome. You've expressed doubt. What do you think might be beyond our reach?

A. Again, I'll use an analogy. We have made great progress in characterizing the human genome. This includes incredibly accurate mapping, or sequencing, of all 3 billion "base pairs" (the genome's alphabet) in the genomes of each of many thousands of people. Yet there are still vast mysteries about how genes are turned on and off to orchestrate the development of healthy cells, organs, and people or what goes wrong in most genetic disorders.

In the brain, there are 80 billion

nerve cells and trillions of connections, or synapses. Accurately mapping all those cells and their connections— what we call the connectome — is currently impossible for the human brain, though it may become feasible for laboratory animals such as fruit flies and even mice. Nonetheless, there are countless exciting discoveries about the human brain still on the horizon.

Q. What should the public understand about this research?

A. There are two reasons to pursue basic research on the brain. The first is human curiosity. How do our brains work? What makes us think, sing, and feel love or hate? Such questions fascinate the lay public as well as neuroscientists. Second, we must pursue this basic research in order to better diagnose, treat, and prevent brain disorders and diseases. All of us have had the experience of seeing someone with a debilitating brain disorder. We need to sustain our efforts to identify cures and provide better treatments.

(Published as "Q&A with David Van Essen," The Galveston County Daily News, *Feb. 5, 2019)*

12. Norman Doidge, March 6, 2019

A remarkable feature of the brain
By William J. Winslade

The human brain is both

magnificent and mysterious. It is extremely complex and protected by our skull. But our brains are very vulnerable to injury from external blows to our head, as well as internal collisions of the brain with the boney internal skull structure.

In addition, disease and deterioration of the brain can cause disability and death. Yet our brains are able to change to recover and be healed.

The final lecture in the Robert and Russell Moody Public Lecture Series will be presented by Norman Doidge, M.D., an internationally renowned psychiatr ist and psychoanalyst.

Dr. Doidge has written two best selling books: *The Brain that Changes Itself* and *The Brain's Way of*

Healing.

In his lecture he will explain how neuroplasticity — the way our brains can change — helps to explain how injured or diseased brains can heal and be restored.

Dr. Doidge will talk about some of his patients who overcame brain disabilities by mobilizing the brain's capacity to change.

This lecture should be of great interest to anyone who wants to learn more about our amazing brain, the most important part of who we are and who we might become.

This lecture series, which began in 2016, has brought together speakers from a wide variety of disciplines. Brain scientists, philosophers, physicians, lawyers, neuroscientists and psychologists have all offered insights about our brains and how they work.

The series has contributed to multi-disciplinary learning. Junior high students, medical students, scientists and the general public have come to the lectures.

This final lecture will provide an even deeper understanding of the way in which our brains can overcome disease or injury.

It is fitting that the last lecture will be given by a remarkable scholar and investigator whose work has helped unravel many mysteries of the brain.

We hope to see you at Moody Gardens IMAX Theater March 6 at 6 p.m.

(Published as "A remarkable author is coming to town," The Galveston County Daily News, *March 2, 2019)*

Appendix 4

Suggestions for further reading

Note: The major books of the lecturers are included in Appendix 1, "Brief Biographies."

Perhaps the best introduction to the mind-body problem is from Rene Descartes himself: *Discourse on Method and the Meditations*: Penguin Books Ltd., 1973.

The most provocative challenge to the generally accepted view of consciousness is Tomas Nagel's *Mind & Cosmos: Why the Materialist Neo-Darwinian Conception of Nature is Almost Certainly False*; Oxford University Press, 2012.

Good questions, good life

E.O. Wilson, *The Meaning of Human Existence*; New York: Liveright, 2015.

Gareth B. Matthews, *Philosophy & the Young Child*; Cambridge, Mass.: Harvard University Press, 1980.

Intellectual histories

As suggested in Chapter 9, intellectual histories, records and autobiographies come in many forms. Here is a sampling.

Ben Franklin, *The Autobiography and Other Writings*; Random House LLC, 2008.

Richard Wright, *Black Boy*; HarperCollins, 1998.

Michel de Montaigne, *The Complete Essays*, translated by M.A. Screech; Penguin Classics, 1993.

Tom Bethel, *Eric Hoffer: The Longshoreman Philosopher*; Hoover

Institution Press, 2012.
Sheilah Graham, *College of One*; Melville House Books, 2013.
Walt Whitman, *Leaves of Grass*; Library of America, 2011.
Antonio Porchia, *Voices*, Spanish text with an English translation by W.S. Merwin; Copper Canyon Press, 2003.

Made in United States
Cleveland, OH
05 November 2026

47669890R10346